The Princeton Review®

Cracking the

AP®

BIOLOGY EXAM

2019 Edition

By the Staff of The Princeton Review

PrincetonReview.com

Penguin
Random
House

The Princeton Review
110 E. 42nd St., 7th Floor
New York, NY 10017
Email: editorialsupport@review.com

Published in the United States by Penguin Random House LLC,
New York, and in Canada by Random House of Canada,
a division of Penguin Random House Ltd., Toronto.

ISBN: 978-1-5247-5796-0
eBook ISBN: 978-1-5247-5830-1
ISSN: 1092-0080

Editor: Sarah Litt
Production Editors: Kathy Carter and Jim Melloan
Production Artist: Craig Patches

Printed in the United States of America on partially
recycled paper.

10 9 8 7 6 5 4 3 2 1

2019 Edition

Editorial
Rob Franek, Editor-in-Chief
Casey Cornelius, Chief Product Officer
Mary Beth Garrick, Director of Production
Selena Coppock, Managing Editor
Meave Shelton, Senior Editor
Colleen Day, Editor
Sarah Litt, Editor
Aaron Riccio, Editor
Orion McBean, Associate Editor

Penguin Random House Publishing Team
Tom Russell, VP, Publisher
Alison Stoltzfus, Publishing Director
Amanda Yee, Associate Managing Editor
Ellen Reed, Production Manager
Suzanne Lee, Designer

Acknowledgments

The Princeton Review would like to thank Katie Chamberlain,
Ph.D, Craig Patches, Jim Melloan, and Kathy Carter for their hard
work on revisions to this edition.

Contents

Get More (Free) Content

1 Go to **PrincetonReview.com/cracking.**

2 Enter the following ISBN for your book: 9781524757960.

3 Answer a few simple questions to set up an exclusive Princeton Review account. (If you already have one, you can just log in.)

4 Click the "Student Tools" button, also found under "My Account" from the top toolbar. You're all set to access your bonus content!

Need to report a potential **content** issue?

Contact **EditorialSupport@review.com**.
Include:

- full title of the book
- ISBN
- page number

Need to report a **technical** issue?

Contact **TPRStudentTech@review.com** and provide:

- your full name
- email address used to register the book
- full book title and ISBN
- computer OS (Mac/PC) and browser (Firefox, Safari, etc.)

The Princeton Review®

Once you've registered, you can...

- Get valuable advice about the college application process, including tips for writing a great essay and where to apply for financial aid

- If you're still choosing between colleges, use our searchable rankings of *The Best 384 Colleges* to find out more information about your dream school.

- Access comprehensive study guides and a variety of printable resources, including bubble sheets and the AP Biology equation and formula tables

- Check to see if there have been any corrections or updates to this edition

- Get our take on any recent or pending updates to the AP Biology Exam

Look For These Icons Throughout The Book

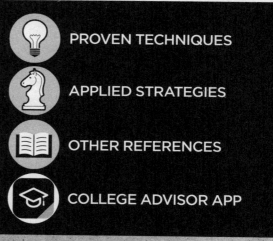

PROVEN TECHNIQUES

APPLIED STRATEGIES

OTHER REFERENCES

COLLEGE ADVISOR APP

Part I
Using This Book to Improve Your AP Score

- Preview: Your Knowledge, Your Expectations
- Your Guide to Using This Book
- How to Begin

PREVIEW: YOUR KNOWLEDGE, YOUR EXPECTATIONS

Welcome to your *Cracking the AP Biology Exam, 2019 Edition*. Your route to a high score on the AP Biology Exam depends a lot on how you plan to use this book. To help you determine your approach, respond to the following questions.

1. Rate your level of confidence about your knowledge of the content tested by the AP Biology Exam.

    A. Very confident—I know it all
    B. I'm pretty confident, but there are topics for which I could use help
    C. Not confident—I need quite a bit of support
    D. I'm not sure

2. If you have a goal score in mind, circle your goal score for the AP Biology Exam.

    5   4   3   2   1   I'm not sure yet

3. What do you expect to learn from this book? Circle all that apply to you.

    A. A general overview of the test and what to expect
    B. Strategies for how to approach the test
    C. The content tested by this exam
    D. I'm not sure yet

YOUR GUIDE TO USING THIS BOOK

Cracking the AP Biology Exam, 2019 Edition is organized to provide as much—or as little—support as you need, so you can use this book in whatever way will be most helpful to improving your score on the AP Biology Exam.

*   The remainder of **Part I** will provide guidance on how to use this book and help you determine your strengths and weaknesses.

*   **Part II** of this book contains Practice Test 1 along with the answers and explanations. (Bubble sheets can be found before each test for easy reference.) We recommend that you take this test before going any further in order to realistically determine:
    o   your starting point right now
    o   which question types you're ready for and which you might need to practice
    o   which content topics you are familiar with and which you will want to carefully review

    Once you have nailed down your strengths and weaknesses with regard to this exam, you can focus your test preparation, build a study plan, and use your time efficiently.

- **Part III** of this book will:
  - provide information about the structure, scoring, and content of the AP Biology Exam
  - help you to make a study plan
  - point you toward additional resources

- **Part IV** of this book will explore:
  - how to attack multiple-choice questions
  - how to write high-scoring free-response answers
  - how to manage your time to maximize the number of points available to you

- **Part V** of this book covers the content you need for your exam.

- **Part VI** of this book contains Practice Test 2 and its answers and explanations. (Again, bubble sheets can be found before each test for easy reference.) Compare your progress between these tests and with Practice Test 1. If you answer a certain type of question wrong several times, you probably need to review it. If you answer it wrong only once, you may have run out of time or been distracted. In either case, comparing your progress will allow you to focus on the factors that caused the discrepancy in scores and to be as prepared as possible on the day of the test.

You may choose to prioritize some parts of this book over others, or you may work through the entire book. Your approach will depend on your needs and how much time you have. Let's now look how to make this determination.

HOW TO BEGIN

1. **Take Practice Test 1**

   Before you can decide how to use this book, you need to take a practice test. Doing so will give you insight into your strengths and weaknesses, and the test will also help you make an effective study plan. If you're feeling test-phobic, remind yourself that a practice test is just a tool for diagnosing yourself—it's not how well you do that matters. As long as you try your best, you can glean invaluable information from your performance to guide your preparation.

   So, before you read further, take Practice Test 1 starting at page 13 of this book. Be sure to finish in one sitting, following the instructions that appear before the test.

2. **Check Your Answers**

   Using the answer key on page 32, count the number of multiple-choice questions you answered correctly and how many you missed. Don't worry about the explanations for now and don't worry about why you missed questions. We'll get to that soon.

3. **Reflect on the Test**

   After you take your first test, respond to the following questions:
   - How much time did you spend on the multiple-choice questions?
   - How much time did you spend on each free-response question?
   - How many multiple-choice questions did you miss?
   - Do you feel you had the knowledge to address the subject matter of the free-response questions?
   - Do you feel you wrote well-organized, thoughtful responses to the free-response questions?

4. **Read Part III of this Book and Complete the Self-Evaluation**

   As discussed in the Guide section above, Part III will provide information on how the test is structured and scored. It will also set out areas of content that are tested.

   As you read Part III, re-evaluate your answers to the questions above. At the end of Part III, you will revisit and refine the questions you answered above. You will then be able to make a study plan, based on your needs and time available, that will allow you to use this book most effectively.

5. **Engage with Parts IV and V as Needed**

   Notice the word *engage*. You'll get more out of this book if you use it intentionally than if you read it passively, hoping for an improved score through osmosis.

   The strategy chapters in Part IV will help you think about your approach to the question types on this exam. Part IV will open with a reminder to think about how you approach questions now and then close with a reflection section asking you to think about how or whether you will change your approach in the future.

   The content chapters in Part V are designed to provide a review of the content tested on the AP Biology Exam, including the level of detail you need to know and how the content is tested. You will have the opportunity to assess your mastery of the content of each chapter through test-appropriate questions.

6. **Take Another Test and Assess Your Performance**

    Once you feel you have developed the strategies you need and gained the knowledge you lacked, you should take Practice Test 2, which starts on page 349 of this book. You should finish in one sitting, following the instructions at the beginning of the test.

    When you are done, check your answers to the multiple-choice sections. See if a teacher will read your essays and provide feedback.

    Once you have taken the test, reflect on what areas you still need to work on and revisit the chapters in this book that address those deficiencies. Through this type of reflection and engagement, you will continue to improve.

7. **Keep Working**

    After you have revisited certain chapters in this book, continue the process of testing, reflection, and engaging with the content in this book and online. Consider what additional work you need to do and how you will change your strategic approach to different parts of the test.

    As we will discuss in Part III, there are other resources available to you, including a wealth of information at AP Students, the official site of the AP Exams. You can continue to explore areas that can stand to improve and engage in those areas right up to the day of the test.

Want More?
For extra practice, check out our supplemental title *550 AP Biology Practice Questions.*

Part II
Practice Test 1

Practice Test 1

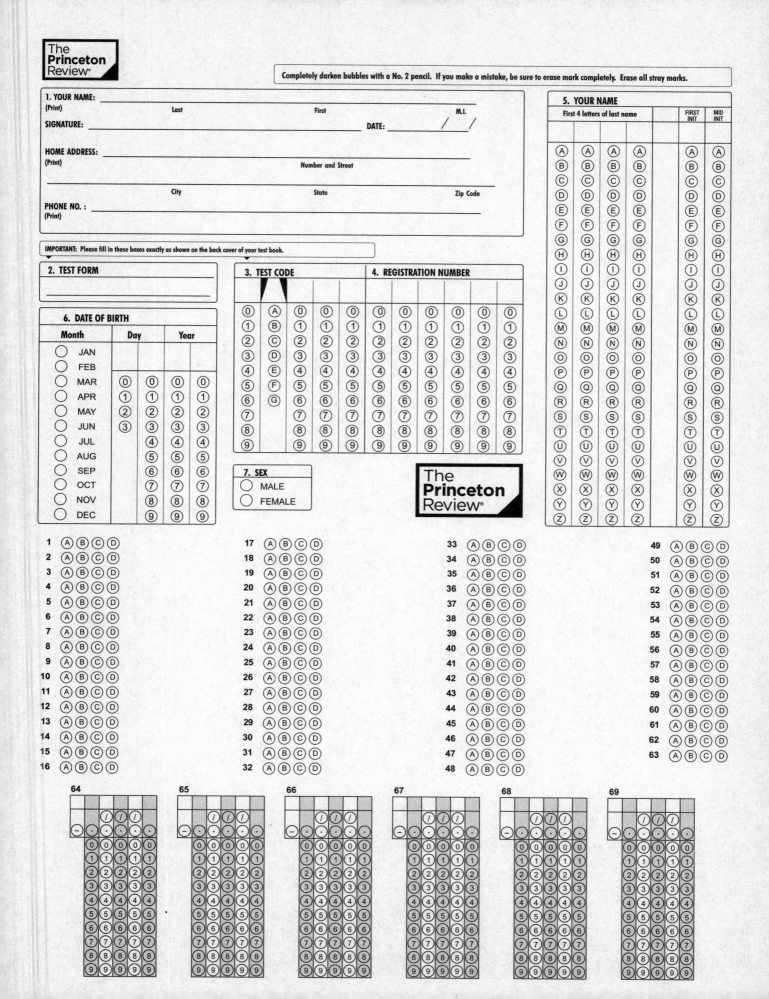

The Exam

AP® Biology Exam

SECTION I: Multiple-Choice Questions

DO NOT OPEN THIS BOOKLET UNTIL YOU ARE TOLD TO DO SO.

At a Glance

Total Time
1 hour and 30 minutes
Number of Questions
69
Percent of Total Score
50%
Writing Instrument
Pencil required

Instructions

Section I of this examination contains 69 multiple-choice questions. These are broken down into Part A (63 multiple-choice questions) and Part B (6 grid-in questions).

Indicate all of your answers to the multiple-choice questions on the answer sheet. No credit will be given for anything written in this exam booklet, but you may use the booklet for notes or scratch work. After you have decided which of the suggested answers is best, completely fill in the corresponding oval on the answer sheet. Give only one answer to each question. If you change an answer, be sure that the previous mark is erased completely. Here is a sample question and answer.

Sample Question   Sample Answer

Chicago is a

(A)   state
(B)   city
(C)   country
(D)   continent

Use your time effectively, working as quickly as you can without losing accuracy. Do not spend too much time on any one question. Go on to other questions and come back to the ones you have not answered if you have time. It is not expected that everyone will know the answers to all the multiple-choice questions.

About Guessing

Many candidates wonder whether or not to guess the answers to questions about which they are not certain. Multiple-choice scores are based on the number of questions answered correctly. Points are not deducted for incorrect answers, and no points are awarded for unanswered questions. Because points are not deducted for incorrect answers, you are encouraged to answer all multiple-choice questions. On any questions you do not know the answer to, you should eliminate as many choices as you can, and then select the best answer among the remaining choices.

BIOLOGY
SECTION I
69 Questions
Time—90 minutes

Directions: Each of the questions or incomplete statements below is followed by four suggested answers or completions. Select the one that is best in each case and then fill in the corresponding oval on the answer sheet.

1. The resting membrane potential depends on which of the following?

   I. Active transport

   II. Selective permeability

   III. Differential distribution of ions across the axonal membrane

   (A) III only
   (B) I and II only
   (C) II and III only
   (D) I, II, and III

2. The Krebs cycle in humans releases

   (A) carbon dioxide
   (B) pyruvate
   (C) glucose
   (D) lactic acid

3. A heterotroph

   (A) obtains its energy from sunlight, harnessed by pigments
   (B) obtains its energy by catabolizing organic molecules
   (C) makes organic molecules from CO_2
   (D) obtains its energy by consuming exclusively autotrophs

4. Regarding meiosis and mitosis, one difference between the two forms of cellular reproduction is that in meiosis

   (A) there is one round of cell division, whereas in mitosis there are two rounds of cell division
   (B) separation of sister chromatids occurs during the second division, whereas in mitosis separation of sister chromatids occurs during the first division
   (C) chromosomes are replicated during interphase, whereas in mitosis chromosomes are replicated during the first phase of mitosis
   (D) spindle fibers form during interphase, whereas in mitosis the spindle fibers form during mitosis

5. A feature of amino acids that is NOT found in carbohydrates is the presence of

   (A) carbon atoms
   (B) oxygen atoms
   (C) nitrogen atoms
   (D) hydrogen atoms

6. Which of the following is NOT a characteristic of bacteria?

   (A) Circular double-stranded DNA
   (B) Membrane-bound cellular organelles
   (C) Plasma membrane consisting of lipids and proteins
   (D) Ribosomes that synthesize polypeptides

7. Which of the following best explains why a population is described as the evolutionary unit?

   (A) Genetic changes can occur only at the population level.
   (B) The gene pool in a population remains fixed over time.
   (C) Natural selection affects individuals, not populations.
   (D) Individuals cannot evolve, but populations can.

8. The endocrine system maintains homeostasis using many feedback mechanisms. Which of the following is an example of positive feedback?

   (A) Infant suckling causes a mother's brain to release oxytocin, which in turn stimulates milk production.
   (B) An enzyme is allosterically inhibited by the product of the reaction it catalyzes.
   (C) When ATP is abundant, the rate of glycolysis decreases.
   (D) When blood sugar levels decrease to normal after a meal, insulin is no longer secreted.

GO ON TO THE NEXT PAGE.

9. A scientist carries out a cross between two guinea pigs, both of which have black coats. Black hair coat is dominant over white hair coat. Three quarters of the offspring have black coats, and one quarter have white coats. The genotypes of the parents were most likely

(A) bb   bb
(B) Bb   Bb
(C) Bb   bb
(D) BB   Bb

10. A large island is devastated by a volcanic eruption. Most of the horses die except for the heaviest males and heaviest females of the group. They survive, reproduce, and perpetuate the population. If weight is a highly heritable trait, which graph represents the change in population before and after the eruption?

(A) A higher mean weight compared with their parents
(B) A lower mean weight compared with their parents
(C) The same mean weight as members of the original population
(D) A higher mean weight compared with members of the original population

11. All of the following play a role in morphogenesis EXCEPT

(A) apoptosis
(B) homeotic genes
(C) operons
(D) differentiation

12. During the period when life is believed to have begun, the atmosphere on primitive Earth contained abundant amounts of all the following gases EXCEPT

(A) oxygen
(B) hydrogen
(C) ammonia
(D) methane

Questions 13–14 refer to the following passage.

The digestive system in humans can be divided into two parts: the alimentary canal and the accessory organs. The canal comprised of the esophagus, stomach, and intestines is where the food actually passes during its transition into waste. The accessory organs are any organs that aid in the digestion by supplying the organs in the alimentary canal with digestive hormones and enzymes.

13. The small intestine is the main site of absorption. It can accomplish it so efficiently because of villi and microvilli that sculpt the membrane into hair-like projections. They likely aid in reabsorption by

(A) increasing the surface area of the small intestine
(B) decreasing the surface area of the small intestine
(C) making the small intestine more hydrophilic
(D) making the small intestine more hydrophobic

14. The pancreas is a major accessory organ in the digestive system. Which of the following would destroy the function of the digestive products produced by the pancreas?

(A) A decrease in absorption rates within the alimentary canal
(B) Removing the excess water from the food waste
(C) Increased acidity due to the inability to neutralize stomach acid
(D) An increase in peristalsis and subsequent diarrhea

15. In animal cells, which of the following represents the most likely pathway that a secreted protein takes as it is synthesized in a cell?

(A) Plasma membrane–Golgi apparatus–ribosome–secretory vesicle–rough ER
(B) Ribosome–Golgi apparatus–rough ER–secretory vesicle–plasma membrane
(C) Plasma membrane–Golgi apparatus–ribosome–secretory vesicle–rough ER
(D) Ribosome–rough ER–Golgi apparatus–secretory vesicle–plasma membrane

16. All of the following statements are correct regarding alleles EXCEPT

(A) alleles are alternative forms of the same gene
(B) alleles are found on corresponding loci of homologous chromosomes
(C) a gene can have more than two alleles
(D) an individual with two identical alleles is said to be heterozygous with respect to that gene

GO ON TO THE NEXT PAGE.

17. Once specific genes, such as the gene coding for ampicillin, have been incorporated into a plasmid, the plasmid may be used to carry out a transformation, which is

(A) inserting it into a bacteriophage
(B) treating it with a restriction enzyme
(C) inserting it into a suitable bacterium
(D) running a gel electrophoresis

18. Although mutations occur at a regular and predictable rate, which of the following statements is the <u>LEAST</u> <u>likely reason the frequency of mutation often</u> *appears* to be low?

(A) Some mutations produce alleles that are recessive and may not be expressed.
(B) Some undesirable phenotypic traits may be prevented from reproducing.
(C) Some mutations cause such drastic phenotypic changes that they are soon removed from the gene pool.
(D) The predictable rate of mutation results in ongoing variability in a gene pool.

19. A scientist wants to test the effect of temperature on seed germination. Which of the following should be part of the experimental design?

(A) Use temperature as the dependent variable and alter the germination times
(B) Use temperature as the independent variable and measure the rate of germination
(C) Use temperature as the controlled variable and keep everything identical between groups
(D) Use the variable natural outside temperature as a control group

20. An autoimmune condition is identified that results in the destruction of Schwann cells. How does this condition impact the lives of patients?

(A) They are unable to regulate their cell cycle and they develop cancer at a high rate.
(B) They are unable to myelinate their neuronal axons and nerve signaling is slow and inconsistent.
(C) They cannot produce helper T-cells and are unable to mount an immune response to even a simple infection.
(D) Their gametes cannot undergo recombination and their offspring will be less genetically variant than they could be.

21. Which of the following is most correct concerning cell differentiation in vertebrates?

(A) Cells in different tissues contain different sets of genes, leading to structural and functional differences.
(B) Differences in the timing and expression levels of different genes lead to structural and functional differences.
(C) Differences in the reading frame of mRNA lead to structural and functional differences.
(D) Differences between tissues result from spontaneous morphogenesis.

<u>Questions 22–23</u> refer to the following passage.

Pumping blood through the human heart must be carefully organized for maximal efficiency and to prevent backflow. In the figure below, the blood enters the heart through the vena cava (1), passes through the right atrium and right ventricle and then goes through the pulmonary artery toward the lungs. After the lungs, the blood returns through the pulmonary vein and then passes into the left atrium and the left ventricle before leaving the heart via the aorta.

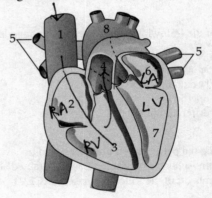

22. Which of the following chambers or vessels carry deoxygenated blood in the human heart?

(A) 1 only
(B) 2 and 3
(C) 1, 2, 3, 4
(D) 4 and 5

23. Blood is pumped via heart contractions triggered by action potentials spreading through the heart muscle. If there is a sudden increase in blood in chamber 3, which chamber of the heart received an increased number of action potentials?

(A) Left atrium
(B) Left ventricle
(C) Right atrium
(D) Right ventricle

GO ON TO THE NEXT PAGE.

24. Some strains of viruses can change normal mammalian cells into cancer cells in vitro. Which of the following is the best explanation for this impact on the mammalian cell?

    (A) A pilus is formed between the mammalian cell and the virus.
    (B) The viral genome incorporates into the mammalian cell's nuclear DNA.
    (C) The host's genome is converted into the viral DNA.
    (D) There is a viral release of spores into the mammalian cell.

25. All of the following correctly describe meiosis EXCEPT

    (A) meiosis produces four haploid gametes
    (B) homologous chromosomes join during synapsis
    (C) sister chromatids separate during meiosis I
    (D) crossing-over increases genetic variation in gametes

26. All of the following are examples of events that can prevent interspecies breeding EXCEPT

    (A) the potential mates experience geographic isolation
    (B) the potential mates experience behavioral isolation
    (C) the potential mates have different courtship rituals
    (D) the potential mates have different alleles

27. Which of the following is NOT a characteristic of asexual reproduction in animals?

    (A) Progeny cells have the same number of chromosomes as the parent cell.
    (B) Progeny cells are identical to the parent cell.
    (C) The parent cell produces diploid cells.
    (D) The progeny cells fuse to form a zygote.

28. Transpiration is a result of special properties of water. The special properties of water include all of the following EXCEPT

    (A) cohesion
    (B) adhesion
    (C) capillary action
    (D) hydrophobicity

Questions 29–32 refer to the following passage.

An experiment was performed to assess the growth of two species of plants when they were grown in different pHs, given different volumes of water, and watered at different times of day over 6 weeks. Two plants were grown of each species and the average heights (in cm) are shown in the table.

		Species A	Species B
pH	2	3.2	4.1
	4	37.6	20.6
	7	62.3	22.4
	10	48.4	31.5
	13	4.1	2.7
Volume (mL)	10	4.9	12.4
	20	19.2	38.9
	40	56.2	45.6
	80	65.1	21.5
	160	2.6	1.8
Time	12:00 A.M.	62.3	20.3
	7:00 A.M.	61.1	21.8
	12:00 P.M.	66.7	18.4
	7:00 P.M.	65.3	19.3

29. For which conditions do the species have different preferences?

    (A) pH
    (B) Volume
    (C) Volume and watering time
    (D) pH and volume and watering time

30. What are the preferred growth conditions for Species B?

    (A) pH 7, 40 mL, any time of day
    (B) pH 10, 40 mL, 7:00 A.M.
    (C) pH 7, 80 mL, any time of day
    (D) pH 10, 80 mL, 12:00 P.M.

31. Which pH and volume were likely used for the watering time experiment?

    (A) pH 4 and 40 mL
    (B) pH 7 and 40 mL
    (C) pH 4 and 80 mL
    (D) pH 7 and 80 mL

GO ON TO THE NEXT PAGE.

32. Which of the following would most improve the statistical significance of the results?

   (A) Let the plants grow for a longer period of time.
   (B) Add more conditions to test, such as amount of light and amount of soil.
   (C) Test the same plants with more pHs and more volumes and times of day.
   (D) Increase the number of plants in each group.

33. Photoperiodism in plants can be best compared to which of the following phenomena in animals?

   (A) Viral infection
   (B) Increased appetite
   (C) Meiotic cell divison
   (D) Circadian rhythms

34. In most ecosystems, net primary productivity is important because it represents the

   (A) energy available to producers
   (B) total solar energy converted to chemical energy by producers
   (C) biomass of all producers
   (D) energy available to heterotrophs

35. Hawkmoths are insects that are similar in appearance and behavior to hummingbirds. Which of the following is LEAST valid?

   (A) These organisms are examples of convergent evolution.
   (B) These organisms were subjected to similar environmental conditions.
   (C) These organisms are genetically related to each other.
   (D) These organisms have analogous structures.

36. Which of the following describes a mutualistic relationship?

   (A) A tapeworm feeds off of its host's nutrients causing the host to lose large amounts of weight.
   (B) Certain plants grow on trees in order to gain access to sunlight, not affecting the tree.
   (C) Remora fish eat parasites off sharks. The sharks stay free of parasites, and the remora fish are protected from predators.
   (D) Meerkats sound alarm calls to warn other meerkats of predators.

37. The pancreas is an organ that makes insulin and glucagon in its beta and alpha cells, respectively. Insulin is released when blood glucose is high and glucagon is released when blood glucose is low. Anti-beta cell antibodies will cause which of the following to occur?

   (A) Glucagon secretion will stop, and blood glucose levels will not decrease.
   (B) Glucagon secretion will stop, and blood glucose levels will decrease.
   (C) Glucagon secretion will stop, and digestive enzymes will be secreted.
   (D) Insulin secretion will stop, and blood glucose levels will not decrease.

Questions 38–40 refer to the following passage.

The rainfall and biomass of several trophic levels in an ecosystem were measured over several years. The results are shown in the graph below.

Rainfall and Biomass 2005–2011

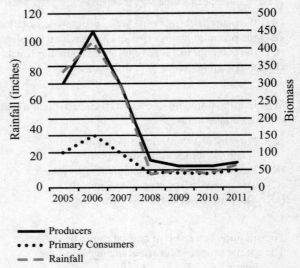

Producers
•••• Primary Consumers
– – Rainfall

38. Which of the following concepts is best demonstrated by this experiment?

   (A) Populations with higher genetic variation can withstand droughts better.
   (B) Meteorological impacts will affect the evolution of populations.
   (C) Environmental changes can affect all the levels of the ecosystem.
   (D) Unoccupied biological niches are dangerous because they attract invasive species.

GO ON TO THE NEXT PAGE.

39. If it rained 120 inches, what would you project the primary consumer biomass to be?

    (A) 150–200
    (B) 60
    (C) 45
    (D) 20

40. Which of the following graphs best depicts the projected biomass of secondary consumers if they were measured?

    (A) **Rainfall and Biomass 2005–2011**

    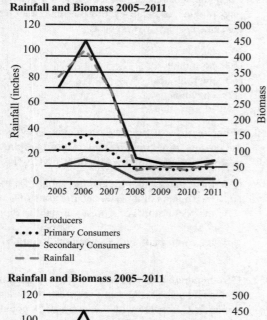

    ——— Producers
    • • • • Primary Consumers
    ——— Secondary Consumers
    – – – Rainfall

    (B) **Rainfall and Biomass 2005–2011**

    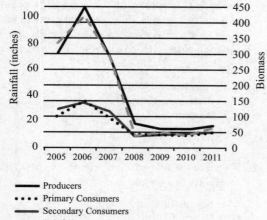

    ——— Producers
    • • • • Primary Consumers
    ——— Secondary Consumers
    – – – Rainfall

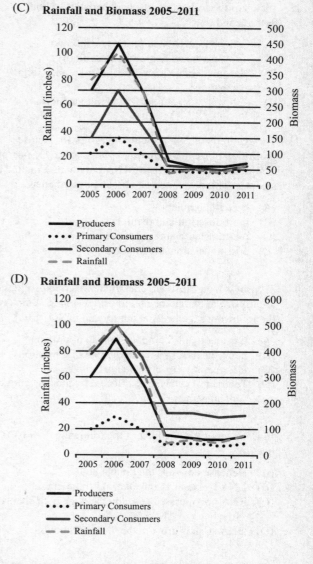

(C) **Rainfall and Biomass 2005–2011**

——— Producers
• • • • Primary Consumers
——— Secondary Consumers
– – – Rainfall

(D) **Rainfall and Biomass 2005–2011**

——— Producers
• • • • Primary Consumers
——— Secondary Consumers
– – – Rainfall

41. The calypso orchid, *Calypso bulbosa*, grows in close association with mycorrhizae fungi. The fungi penetrate the roots of the flower and take advantage of the plant's food resources. The fungi concentrate rare minerals, such as phosphates, in the roots and make them readily accessible to the orchid. This situation is an example of

    (A) parasitism
    (B) commensalism
    (C) mutualism
    (D) endosymbiosis

GO ON TO THE NEXT PAGE.

42. Which of the following are characteristics of both bacteria and fungi?

   (A) Cell wall, DNA, and plasma membrane
   (B) Nucleus, organelles, and unicellularity
   (C) Plasma membrane, multicellularity, and Golgi apparatus
   (D) Cell wall, unicellularity, and mitochondria

43. The synthesis of new proteins necessary for lactose utilization by the bacterium *E. coli* using the *lac* operon is regulated by the ability of RNA polymerase to bind and advance. This regulation can best be described as

   (A) bacterial regulation
   (B) pre-transcriptional regulation
   (C) pre-translational regulation
   (D) post-translational regulation

44. Trypsin is a digestive enzyme. It cleaves polypeptides after lysine and arginine amino acid residues. Which of the following statements about trypsin is NOT true?

   (A) It is an organic compound made of proteins.
   (B) It is a catalyst that alters the rate of a reaction.
   (C) It is operative over a wide pH range.
   (D) The rate of catalysis is affected by the concentration of substrate.

45. In DNA replication, which of the following does NOT occur?

   (A) Helicase unwinds the double helix.
   (B) DNA ligase links the Okazaki fragments.
   (C) RNA polymerase is used to elongate both chains of the helix.
   (D) DNA strands grow in the 5′ to 3′ direction.

46. A change in a neuron membrane potential from +50 millivolts to –70 millivolts is considered

   (A) depolarization
   (B) repolarization
   (C) hyperpolarization
   (D) an action potential

47. The energy given up by electrons as they move through the electron transport chain is used to

   (A) break down glucose
   (B) make glucose
   (C) produce ATP
   (D) make NADH

48. If a plant undergoing the light-dependent reactions of photosynthesis began to release $^{18}O_2$ instead of normal oxygen, one could most reasonably conclude that the plant had been supplied with

   (A) H_2O containing radioactive oxygen
   (B) CO_2 containing radioactive oxygen
   (C) $C_6H_{12}O_6$ containing radioactive oxygen
   (D) NO_2 containing radioactive oxygen

49. Chemical substances released by organisms that elicit a physiological or behavioral response in other members of the same species are known as

   (A) auxins
   (B) hormones
   (C) pheromones
   (D) enzymes

50. Homologous structures are often cited as evidence for the process of natural selection. All of the following are examples of homologous structures EXCEPT

   (A) the forearms of a cat and the wings of a bat
   (B) the flippers of a whale and the arms of a man
   (C) the pectoral fins of a porpoise and the flippers of a seal
   (D) the forelegs of an insect and the forelimbs of a dog

51. Certain populations of finches have long been isolated on the Galapagos Islands off the western coast of South America. Compared with the larger stock population of mainland finches, these separate populations exhibit far greater variation over a wider range of species. The variation among these numerous finch species is the result of

   (A) convergent evolution
   (B) divergent evolution
   (C) disruptive selection
   (D) stabilizing selection

52. Which of the following contributes the MOST to genetic variability in a population?

   (A) Sporulation
   (B) Binary fission
   (C) Vegetative propagation
   (D) Mutation

Questions 53–55 refer to the following information and table.

A marine ecosystem was sampled in order to determine its food chain. The results of the study are shown below.

Type of Organism	Number of Organisms
Shark	2
Small crustaceans	400
Mackerel	20
Phytoplankton	1,000
Herring	100

53. Which of the following organisms in this population are secondary consumers?

(A) Sharks
(B) Phytoplankton
(C) Herrings
(D) Small crustaceans

54. Which of the following organisms has the largest biomass in this food chain?

(A) Phytoplanktons
(B) Mackerels
(C) Herrings
(D) Sharks

55. If the herring population is reduced by predation, which of the following would most likely be a secondary effect on the ecosystem?

(A) The mackerels will be the largest predator in the ecosystem.
(B) The small crustacean population will be greatly reduced.
(C) The phytoplankton population will be reduced over the next year.
(D) The small crustaceans will become extinct.

Questions 56–58 refer to the following information and diagram.

To understand the workings of neurons, an experiment was conducted to study the neural pathway of a reflex arc in frogs. A diagram of a reflex arc is given below.

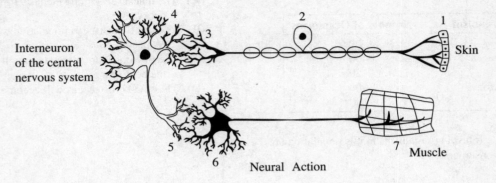

56. Which of the following represents the correct pathway taken by a nerve impulse as it travels from the spinal cord to effector cells?

(A) 1-2-3-4
(B) 6-5-4-3
(C) 2-3-4-5
(D) 4-5-6-7

57. The brain of the frog is destroyed. A piece of acid-soaked paper is applied to the frog's skin. Every time the piece of paper is placed on its skin, one leg moves upward. Which of the following conclusions is best supported by the experiment?

(A) Reflex actions are not automatic.
(B) Some reflex actions can be inhibited.
(C) All behaviors in frogs are primarily reflex responses.
(D) This reflex action does not require the brain.

58. A nerve impulse requires the release of neurotransmitters at the axonal bulb of a presynaptic neuron. Which of the following best explains the purpose of neurotransmitters, such as acetylcholine?

(A) They speed up the nerve conduction in a neuron.
(B) They open the sodium channels in the axonal membrane.
(C) They excite or inhibit the postsynaptic neuron.
(D) They open the potassium channels in the axonal membrane.

GO ON TO THE NEXT PAGE.

Questions 59–61 refer to the figure and chart below.

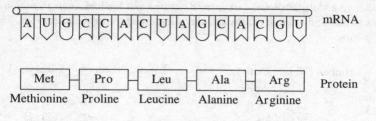

Formation of a Protein

		The Genetic Code: Codons of mRNA that Specify a Given Amino Acid			
First Position (5' end)	**Second Position**	**Third Position (3' end)**			
		U	**C**	**A**	**G**
U	U	UUU UUC Phenylalanine		UUA UUG Lucine	
	C	UCU UCC UCA UCG Serine			
	A	UAU UAC Tytrosine		UAA UAG	
	G	UGU UGC Cysteine		UGA	UGG Tryptophan
C	U	CUU CUC CUA CUG Leucine			
	C	CCU CCC CCA CCG Proline			
	A	CAU CAC Histidine		CAA CAG Glutamine	
	G	CGU CGC CGA CGG Arginine			
A	U	AUU AUC AUA Isoleucine			AUG
	C	ACU ACC ACA ACG Threonine			
	A	AAU AAC Asparagine		AAA AAG Lysine	
	G	AGU AGC Serine		AGA AGG Arginine	
G	U	GUU GUC GUA GUG Valine			
	C	GCU GCC GCA GCG			
	A	GAU GAC Aspartic Acid		GAA GAG Glutamic acid	
	G	GGU GGC GGA GGG Glycine			

59. Which of the following DNA strands is the template strand that led to the amino acid sequence shown above?

(A) 3'-ATGCGACCAGCACGT-5'
(B) 3'-AUGCCACUAGCACGU-5'
(C) 3'-TACGGTGATCGTGCA-5'
(D) 3'-UACGGUGAUCGUGCA-5'

60. Immediately after the translation of methionine, a chemical is added which deletes all remaining uracil nucleotides in the mRNA. Which of the following represents the resulting amino acid sequence?

(A) Serine–histidine–serine–threonine
(B) Methionine–proline–glutamine–histidine
(C) Methionine–proline–leucine–alanine–arginine
(D) Methionine–proline–alanine–arginine–arginine

61. The mRNA above was found to be much smaller than the original mRNA synthesized in the nucleus. This is due to the

(A) addition of a poly(A) tail to the mRNA molecule
(B) addition of a cap to the mRNA molecule
(C) excision of exons from the mRNA molecule
(D) excision of introns from the mRNA molecule

GO ON TO THE NEXT PAGE.

Questions 62 and 63 refer to the following information.

A scientist studies the storage and distribution of oxygen in humans and Weddell seals to examine the physiological adaptations that permit seals to descend to great depths and stay submerged for extended periods. The figure below depicts the oxygen storage in both organisms.

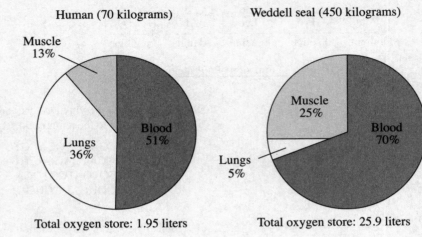

Human (70 kilograms)   Weddell seal (450 kilograms)

Total oxygen store: 1.95 liters   Total oxygen store: 25.9 liters

62. Compared with humans, approximately how many liters of oxygen does the Weddell seal store per kilogram of body weight?

(A) The same amount of oxygen
(B) Twice the amount of oxygen
(C) Three times the amount of oxygen
(D) Five times the amount of oxygen

63. During a dive, a Weddell seal's blood flow to the abdominal organs is shut off, and oxygen-rich blood is diverted to the eyes, brain, and spinal cord. Which of the following is the most likely reason for this adaptation?

(A) To increase the number of red blood cells in the nervous system
(B) To increase the amount of oxygen reaching the skeletomuscular system
(C) To increase the amount of oxygen reaching the central nervous system
(D) To increase the oxygen concentration in the lungs

GO ON TO THE NEXT PAGE.

Directions: Part B consists of questions requiring numeric answers. Calculate the correct answer for each question.

64. An experiment was conducted to observe the light-absorbing properties of chlorophylls and carotenoids using a spectrophotometer. The pigments were first extracted and dissolved in a solution. They were then illuminated with pure light of different wavelengths to detect which wavelengths were absorbed by the solution. The results are presented in the absorption spectrum below.

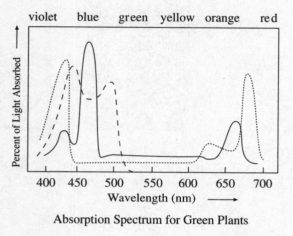

Absorption Spectrum for Green Plants

·············· Chlorophyll *a*
———— Chlorophyll *b*
– – – – Carotenoids

At approximately what wavelength does chlorophyll *a* maximally absorb light?

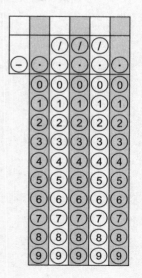

65. A woman with blood genotype $I^A i$ and a man with blood genotype $I^B i$ have two children, both type AB. What is the probability that a third child will be blood type AB?

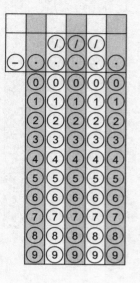

GO ON TO THE NEXT PAGE.

66. The trophic level efficiency of large herbivores such as elks is frequently only about 5 percent. In tons, what volume of plants would be required to maintain 24,000 lbs of elk?

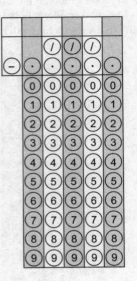

67. If the genotype frequencies of an insect population are AA = 0.49, Aa = 0.42, and aa = 0.09, what is the gene frequency of the recessive allele?

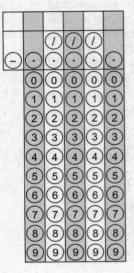

GO ON TO THE NEXT PAGE.

Question 68 refers to the diagram below.

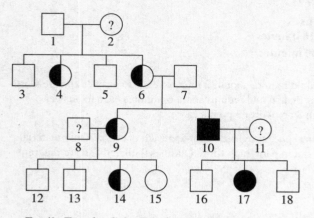

Family Tree for Color Blindness: squares represent males and circles represent females. Shading in a shape represents the allele for color blindness. Each half of a shape represents one of the two copies of the gene.

68. Based on the pedigree above, what is the probability that a male child born to individuals 6 and 7 will be color-blind?

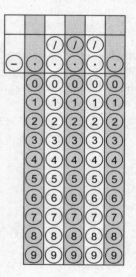

69. The loss of water by evaporation from the leaf openings is known as transpiration. The transpiration rates of various plants are shown below.

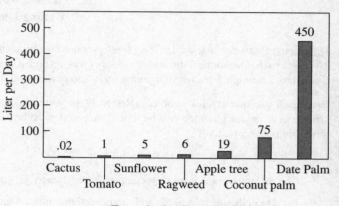

Transpiration Rates for Plants

How many liters of water per week are lost by a coconut palm?

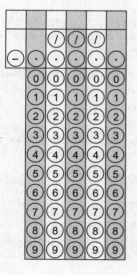

STOP

END OF SECTION I

IF YOU FINISH BEFORE TIME IS CALLED, YOU MAY CHECK YOUR WORK ON THIS SECTION. DO NOT GO ON TO SECTION II UNTIL YOU ARE TOLD TO DO SO.

BIOLOGY
SECTION II
8 Questions
Planning Time—10 minutes
Writing Time—80 minutes

Directions: Questions 1 and 2 are long free-response questions that should require about 22 minutes each to answer and are worth 10 points each. Questions 3 through 8 are short free-response questions that should require about 6 minutes each to answer. Questions 3 through 5 are worth 4 points each, and questions 6 through 8 are worth 3 points each.

Read each question carefully and completely. Write your response in the space provided following each question. Only material written in the space provided will be scored. Answers must be written out in paragraph form. Outlines, bulleted lists, or diagrams alone are not acceptable.

1. The cell membrane is an important structural feature of a nerve cell.

   (a) **Describe** the role of the sodium potassium pump in maintaining the resting membrane potential.

   (b) **Discuss** ion flow during an action potential.

   (c) **Predict** the outcome on action potentials if a cell could not make voltage-gated sodium channels.

   (d) **Explain** how myelin affects the speed of an action potential.

2. Sickle-cell anemia is a genetic disorder caused by the abnormal gene for hemoglobin S. A single substitution occurs in which glutamic acid is substituted for valine in the sixth position of the hemoglobin molecule. This change reduces hemoglobin's ability to carry oxygen.

   (a) **Discuss** the process by which mutation occurs in base substitution.

   (b) Biologists used gel electrophoresis to initially identify the mutant gene. **Explain** how gel electrophoresis could be applied to the identification of the gene mutation. **Discuss** the use of restriction enzymes.

   (c) Hemoglobin S is transmitted as a simple Mendelian allele. **Describe** the outcome if a female who does not carry the abnormal allele mates with a male homozygous for the disease. **Include** phenotypic and genotypic ratios.

3. Cell size is limited by the surface area-to-volume ratio of the cell membrane.

   (a) **Discuss** why cell size is limited by this ratio.

   (b) **Describe** two adaptations that increase surface area in organisms.

   (c) **Describe** the difference in how small polar and small nonpolar molecules cross cell membranes

4. **Discuss** the Krebs cycle, the electron transport chain, and chemiosmosis.

   (a) **Explain** why these steps are considered aerobic processes.

   (b) **Discuss** the location at which **each** stage occurs.

GO ON TO THE NEXT PAGE.

5. __Describe__ three main differences between meiosis and mitosis.

6. __Define__ homologous structures and give an example. __Describe__ how they are different from analogous structures.

7. __Describe__ the three ways that genetic information is transmitted laterally between bacteria.

8. __Describe__ why viruses are typically not considered to be alive.

STOP

END OF EXAM

Practice Test 1:
Answers and
Explanations

PRACTICE TEST 1 ANSWER KEY

1.	D	24.	B	47.	C
2.	A	25.	C	48.	A
3.	B	26.	D	49.	C
4.	B	27.	D	50.	D
5.	C	28.	D	51.	B
6.	B	29.	B	52.	D
7.	D	30.	A	53.	C
8.	A	31.	D	54.	A
9.	B	32.	D	55.	C
10.	D	33.	D	56.	D
11.	C	34.	D	57.	D
12.	A	35.	C	58.	C
13.	A	36.	C	59.	C
14.	C	37.	D	60.	B
15.	D	38.	C	61.	D
16.	D	39.	A	62.	B
17.	C	40.	A	63.	C
18.	D	41.	C	64.	425–440
19.	B	42.	A	65.	1/4 or 0.25 or 25%
20.	B	43.	B	66.	240
21.	B	44.	C	67.	0.3 or 30%
22.	C	45.	C	68.	1/2 or 0.5 or 50%
23.	C	46.	B	69.	525

PRACTICE TEST 1 EXPLANATIONS

Section I: Multiple Choice

1. **D** The resting potential depends on active transport (the Na$^+$K$^+$-ATPase pump) and the selective permeability of the axon membrane to K$^+$ than to Na$^+$, which leads to a differential distribution of ions across the axonal membrane.

2. **A** The Krebs cycle releases carbon dioxide as the carbon molecules are broken down and electron carriers are generated. Glucose is at the start of glycolysis and pyruvate is at the end of glycolysis. Lactic acid is generated through fermentation.

3. **B** A heterotroph obtains its energy from organic molecules. An autotroph obtains energy from sunlight utilizing pigments such as chlorophyll and uses CO_2 and water to make organic molecules. Therefore, (A) and (C) can be eliminated. Heterotrophs can obtain their energy from ingesting autotrophs, but they can also consume other heterotrophs. So you can eliminate (D), leaving (B) as your answer.

4. **B** In meiosis, the sister chromatids separate during the second metaphase of meiosis (Meiosis II), whereas the sister chromatids separate during metaphase of mitosis. Choice (A) is incorrect because in meiosis there are two rounds of cell division, whereas in mitosis there is only one round of cell division. Chromosomes are replicated during interphase in both meiosis and mitosis, so (C) is incorrect. Choice (D) can also be eliminated because spindle fibers never form during interphase.

5. **C** Amino acids are organic molecules that contain carbon, hydrogen, oxygen, and nitrogen, so eliminate (A), (B), and (D). Don't forget to associate amino acids with nitrogen because of the amino group (NH_2).

6. **B** Unlike eukaryotes, prokaryotes (which include bacteria) do not contain membrane-bound organelles. Bacteria contain circular double-stranded DNA, ribosomes, and a cell wall, so (A) and (D) are incorrect. Also eliminate (C) because bacterial cell membranes are made up of a bilipid layer with proteins interspersed.

7. **D** Populations can be described as the evolutionary unit because changes in the genetic makeup of populations can be measured over time. Eliminate (A), as genetic changes occur only at the individual level. Only under Hardy-Weinberg equilibrium does the gene pool remain fixed over time in a population. However, this statement does not explain why the population is the evolving unit, so (B) is incorrect. Choice (C) is true but does not address the question.

8. **A** Positive feedback occurs when a stimulus causes an increased response. Choices (B), (C), and (D) are examples of negative feedback.

9. **B** In order to determine the genotype of the parents, use the ratio of the offspring given in the question and work backward. The ratio of black-haired to white-haired guinea pigs is 3:1. In order to get a white-haired offspring, each parent must have been able to contribute a b allele. However, since the offspring were both black in color, they must have each been Bb.

10. **D** The mean weight of the offspring in the next generation will be heavier than the mean weight of the original population because all the lighter horses in the original population died off. The normal distribution for weight will therefore shift to the heavier end (to the right of the graph). You can therefore eliminate (C) because the mean weight should increase. The mean weight of the offspring could be heavier or lighter than their parents, so you can also eliminate (A) and (B).

11. **C** Apoptosis (programmed cell death), hox and homeotic genes (genes that control differentiation), and differentiation itself play a role in morphogenesis. Operons are sets of multiple genes regulated by a single regulatory unit in bacteria.

12. **A** The primitive atmosphere lacked oxygen (O_2). It contained methane (CH_4), ammonia (NH_3), and hydrogen (H_2).

13. **A** Villi and microvilli are fingerlike projections present in the small intestine which dramatically increase the surface area available for nutrient absorption.

14. **C** The pancreas produces digestive enzymes, and enzymes are sensitive to pH. The inability to neutralize stomach acid would disrupt the function of the enzymes. Food moving too quickly would decrease digestion/absorption rates but the function of the enzyme would be intact. Removal of water should not greatly affect enzyme function.

15. **D** Ribosomes are the site of protein synthesis. Therefore, the correct answer should start with ribosome. So eliminate (A) and (C). The polypeptide then moves through the rough ER to the Golgi apparatus, where it is modified and packaged into a vesicle. The vesicle then floats to the plasma membrane and is secreted. Choice (D) is your answer.

16. **D** This statement is false because an individual with two identical alleles is said to be *homozygous,* not *heterozygous,* with respect to that gene. Alleles are different forms of the same gene found on corresponding positions of homologous chromosomes, so (A) and (B) are incorrect. More than two alleles can exist for a gene, but a person can have only two alleles for each trait.

17. **C** A transformation is the uptake of DNA by a bacterium. DNA uptake by a virus (A) would be a transduction, although this requires a virus to gain the DNA via infection of a bacterium.

18. **D** The least likely explanation for why mutations are low is that mutations produce variability in a gene pool. Any gene is bound to mutate. This produces a constant input of new genetic information into a gene pool. This answer choice doesn't give us any additional information about the rate of mutations. Some mutations are subtle and cause only a slight decrease in reproductive output, so eliminate (A). Some mutations are harmful and decrease the productive success of the individual, so (B) is incorrect. Some mutations are deleterious and lead to total reproductive failure. The zygote fails to develop. Therefore, (C) can also be eliminated.

19. **B** The independent variable is the condition the scientist sets up. The dependent variable is the thing that is the measured outcome of the experiment. The scientist should set different temperatures and then measure the germination rate. The outdoor temperature is interesting, but it is too variable to be a good control for the groups with a set temperature.

20. **B** Schwann cells are responsible for the production of myelin. Without them, axons will be unmyelinated and will be unable to function properly.

21. **B** Every cell in an organism has the same set of genes. Differences in the timing and expression levels of these genes lead to structural and functional differences.

22. **C** Blood flows in all chambers before the lungs is deoxygenated. This means the correct answer is chambers 1, 2, 3, and 4.

23. **C** The action potentials cause the heart to contract. If chamber 3 is full of more blood, then that would indicate that the preceding chamber, chamber 2 (right atrium), had suddenly contracted a larger amount.

24. **B** Normal cells can become cancerous when a virus invades the cell and takes over the replicative machinery. This would occur if the mammalian genome is altered, such as if the viral genome inserted itself and disrupted mammalian gene expression. A pilus forms between two bacteria, so (A) is wrong. Also eliminate (C) because the host's genome is not converted to the viral genome. Choice (D) is incorrect because spores are released by fungi, not viruses.

25. **C** Crossing-over and synapsis occur during meiosis, which produces haploid gametes. Separation of homologous chromosomes occurs during meiosis I, while separation of sister chromatids does not occur until meiosis II.

26. **D** If potential mates have different alleles, this doesn't keep them from mating. Having different alleles of a gene is just a normal part of genetic variation. Use common sense to eliminate the other answer choices. If the organisms don't meet, they won't reproduce; eliminate (A). Also eliminate (B) and (C): if potential mates do not share the same behaviors (such as courtship rituals), they may not mate.

27. **D** There is no union of gametes in mitosis. Choices (A) and (C) are incorrect: asexual reproduction involves the production of two new cells with the same number of chromosomes as the parent cell. If the parent cell is diploid, then the daughter cells will be diploid. The daughter cells are identical to the parent cell, so eliminate (B).

28. **D** Cohesion, adhesion, and capillary action are special properties of water that are necessary for transpiration to occur. Water is a hydrophilic, polar molecule.

29. **B** The two species differ in their preference for water volume. Species A prefers 80 mL, and Species B prefers 40 mL. The different watering times do not seem to play a role in growth as there is no obvious pattern and all of the times had similar growth. Both plants prefer pH 7.

30. **A** The top growth conditions for Species B are pH 7, 40 mL, and any time of day. In those conditions, the Species B plants grew the tallest.

31. **D** The plants in the watering time experiment seemed to be around 60 and 20 cm in height. This corresponds to a pH of 7 rather than 4 and a watering volume of 80 mL rather than 40 mL.

32. **D** Choices (A), (B), and (C) would all give interesting information about the experiment, but the only one that would make the results more statistically significant is to increase the number plants in each group. Two plants is not enough to be sure about the results of the experiment.

33. **D** Photoperiodism is a response of plants to the changing length of day/night. It is a way that plants respond to a signal from the environment. Animal circadian rhythms are also a response to the environment (D). Viral infection (A), increased appetite (B), and meiotic division (C) are not responses to the environment.

34. **D** The net primary productivity is what the energy producers have left for storage after their own energy needs have been met. Energy stored = (total energy produced) − (energy used in cellular respiration). Choice (A) is incorrect; it is not the energy available to producers. You can also eliminate (B) because it is not the total chemical energy in producers—that is the gross primary productivity. Get rid of (C) as well because it is not the biomass (total living material) among producers.

35. **C** Choices (A) and (B) are incorrect: these organisms exhibit the same behavior because they were subjected to the same environmental conditions and similar habitats. This is an example of convergent evolution. However, they are not genetically similar, so (C) is the answer. (One is an insect, and the other a bird.) They are analogous, so (D) is also incorrect. They exhibit the same function but are structurally different.

36. **C** A mutualistic relationship is a relationship among two organisms in which both benefit. Choice (A) describes parasitism, and (B) describes commensalism. Choice (D) is an example of altruistic behavior.

37. **D** Beta cells secrete insulin. Destruction of beta cells in the pancreas will halt the production of insulin. Therefore, eliminate (A), (B), and (C). This will lead to an increase in blood glucose levels.

38. **C** The graph does show a drought, but it says nothing about genetic variation. It is true that the rainfall could affect evolution of the population, but the graph doesn't address evolution. It shows only the decrease in biomass. Choice (C) is the best answer since it addresses how the environment effects ripple through the levels of the ecosystem. Invasive species do not need an unoccupied niche, and that is not shown in the graph.

39. **A** The biomass seems to correlate fairly well with the rainfall. Therefore, a rainfall higher than any recorded would give a biomass higher than any recorded. The axis on the right shows the biomass. With a rainfall of 120, the biomass should be greater than 150.

40. **A** The secondary consumers' biomass should always be less than that of the primary consumers, so the only option is (A).

41. **C** This is an example of mutualism. Both organisms benefit. Choice (A), parasitism, is a type of symbiotic relationship in which one organism benefits and the other is harmed. Choice (B), commensalism, occurs when one organism benefits and the other is unaffected. Choice (D), endosymbiosis, is the idea that some organelles originated as symbiotic prokaryotes that live inside larger cells.

42. **A** Both bacteria and fungi contain genetic material (DNA), a plasma membrane, and a cell wall. Unlike fungi, bacteria lack a definite nucleus. Therefore, eliminate (B). Bacteria are unicellular, whereas fungi are both unicellular and multicellular. Therefore, eliminate (C) and (D).

43. **B** This question tests your understanding of what stage of gene expression utilizes RNA polymerase. Since the presence of RNA pol is required for transcription, the regulation is pre-transcriptional.

44. **C** Trypsin is an enzyme. Enzymes are proteins and organic catalysts that speed up reactions without altering them. They are not consumed in the process. Therefore, you can eliminate (A) and (B). The rate of reaction can be affected by the concentration of the substrate up to a point, so (D) can be eliminated.

45. **C** DNA polymerase, not RNA polymerase, is the enzyme that causes the DNA strands to elongate. DNA helicase unwinds the double helix, so (A) is true and therefore incorrect. Choice (B), which states that DNA ligase seals the discontinuous Okazaki fragments, is also true. Eliminate it. In the presence of DNA polymerase, DNA strands always grow in the 5′ to 3′ direction as complementary bases attach. Therefore, (D) is also incorrect.

46. **B** A voltage change from +50 to −70 is called repolarization. You can eliminate (A) because a voltage change from −70 to +50 is called depolarization. Choice (C) is incorrect; a voltage change from −70 to −90 is called hyperpolarization. Choice (D) can be eliminated as well; an action potential is a traveling depolarized wave. It refers to the whole thing, from depolarization, repolarization, hyperpolarization, and back to a resting potential.

47. **C** Electrons passed down along the electron transport chain from one carrier to another lose energy and provide energy for making ATP. Glucose is decomposed during glycolysis, but this process is not associated with energy given up by electrons; eliminate (A). Glucose is made during photosynthesis, so eliminate (B). NADH is an energy-rich molecule, which accepts electrons during the Krebs cycle. Therefore, (D) is incorrect as well.

48. **A** The oxygen released during the light reaction comes from splitting water. (Review the reaction for photosynthesis.) Therefore, water must have originally contained the radioactive oxygen. Carbon dioxide is involved in the dark reaction and produces glucose, so eliminate (B). Glucose is the final product and would not be radioactive unless carbon dioxide was the radioactive material, so (C) is incorrect. Finally, eliminate (D), because nitrogen is not part of photosynthesis.

49. **C** Pheromones act as sex attractants, alarm signals, or territorial markers. Auxins are plant hormones that promote growth, so eliminate (A). Hormones are chemical messengers that produce a specific effect on target cells within the same organism, so eliminate (B). Enzymes are catalysts that speed up reactions, so eliminate (D).

50. **D** Homologous structures are organisms with the same structure but different functions. The forelegs of an insect and the forelimbs of a dog are not structurally similar. (One is an invertebrate, and the other a vertebrate.) They do not share a common ancestor. However, both structures are used for movement. All of the other examples are vertebrates that are structurally similar.

51. **B** Speciation occurred in the Galapagos finches as a result of the different environments on the islands. This is an example of divergent evolution. The finches were geographically isolated. Choice (A), convergent evolution, is the evolution of similar structures in distantly related organisms. Choice (C), disruptive selection, is selection that favors both extremes at the expense of the intermediates in a population. Choice (D), stabilizing selection, is selection that favors the intermediates at the expense of the extreme phenotypes in a population.

52. **D** Mutations produce genetic variability. All of the other answer choices are forms of asexual reproduction.

53. **C** Secondary consumers feed on primary consumers. If you set up a pyramid of numbers, you'll see that the herrings belong to the third trophic level.

54. **A** The biomass is the total bulk of a particular living organism. The phytoplankton population has both the largest biomass and the most energy.

55. **C** If the herring population decreases, this will lead to an increase in the number of crustaceans and a decrease in the phytoplankton population. Reorder the organisms according to their trophic levels and determine which populations will increase and decrease accordingly.

56. **D** This question tests your ability to trace the neural pathway of a motor (effector) neuron. The nerve conduction will travel from the spinal cord (where interneurons are located) to the muscle.

57. **D** Because the brain is destroyed, it is not associated with the movement of the leg. Choice (A) is incorrect, as reflex actions are automatic. Choices (B) and (C) can also be eliminated; both of these statements are true but are not supported by the experiment.

58. **C** Neurotransmitters are released from the axonal bulb of one neuron and diffuse across a synapse to activate a second neuron. The second neuron is called a postsynaptic neuron. A neurotransmitter can either excite or inhibit the postsynaptic neuron. The myelin sheath speeds up the conduction in a neuron, so (A) is wrong. Also eliminate (B) and (D), as both sodium and potassium channels open during an action potential. Neurotransmitters are not involved in actions related to the axon membrane. They do not force potassium ions to move against a concentration gradient.

59. **C** The DNA template strand is complementary to the mRNA strand. Using the mRNA strand, work backward to establish the sequence of the DNA strand. Don't forget that DNA strands do not contain uracil, so eliminate (B) and (D).

60. **B** Use the amino acid chart to determine the sequence after uracil is deleted. The deletion of uracil creates a frameshift.

61. **D** The mRNA is modified before it leaves the nucleus. It becomes smaller when introns (intervening sequences) are removed. A poly(A) tail and a cap are added to the mRNA and would therefore increase the length of the mRNA, so you can eliminate (A) and (B). Choice (C) is also incorrect, as exons are the coding sequences that are kept by the mRNA.

62. **B** The Weddell seal stores twice as much oxygen as humans. Calculate the liters per kilograms weight for both the seal and man using the information at the bottom of the chart. The Weddell seal stores 0.058 liters/kilograms (25.9 liters/450 kilograms) compared to 0.028 liters/kilograms (1.95 liters/70 kilograms) in humans.

63. **C** The most plausible answer is that blood is redirected toward the central nervous system, which permits the seal to navigate for long durations. Choice (A) is incorrect; the seal does not need to increase the number of red blood cells in the nervous system. Choice (B) can also be eliminated, as the seal does not need to increase the amount of oxygen to the skeletal system. Eliminate (D) because the diversion of blood does not increase the concentration of oxygen in the lungs.

64. **425–450**

    If you look at the absorption spectrum, you'll see that chlorophyll *a* has two peaks, one at 425 nm and one at 680 nm. Chlorophyll *a* maximally absorbs light at approximately 425 nm.

65. $\frac{1}{4}$ or **0.25** or **25%**

    Make a Punnett square to determine the probability that the couple has a child with blood type AB. The probability is $\frac{1}{4}$ whether it's the first child or the third child.

66. **240** 24,000 lbs of elk is equivalent to 12 tons since there are 2,000 lbs per ton. With a 5 percent efficiency of transfer from their food source, that 12 tons represents 5 percent of the weight of the plants they would need to consume. 5/100 = 12/tons needed, making the answer 240.

67. **0.3** or **30%**

    The frequency of the homozygous dominant genotype (AA) is 0.49. To find the dominant allele frequency, we can use the formula provided by the Hardy-Weinberg theory, $p^2 + 2pq + q^2 = 1$, where p represents the dominant allele and q represents the recessive allele. Because we know that p^2 represents the frequency of the homozygous dominant genotype, we can find the frequency of the dominant allele (p) by taking the square root of the frequency of the genotype. The square root of 0.49 is 0.7. Note that by using the formula $p + q = 1$, we can also determine the frequency of the recessive allele (q). It is 0.3 ($0.7 + q = 1$).

68. $\frac{1}{2}$ or **0.5** or **50%**

    Draw a Punnett square for the couple to determine the probability of color blindness for the boys.

    Individuals 6 and 7 (X^cX and XY) will produce two males: one X^cY and one XY. The probability of color blindness is therefore $\frac{1}{2}$.

69. **525** Make sure you read the question carefully. You are asked to calculate the number of liters per week, not per day. The chart tells us that a coconut palm loses 75 liters a day, which would mean 525 liters a week ($7 \times 75 = 525$).

Section II: Free Response

On the following pages, you'll find two types of aids for grading your essays: checklists and sample responses. Checklists like these are used to grade your essays on the actual test. They're actually quite simple to use.

For each item you mentioned, give yourself the appropriate number of points (1 point, 2 points, and so on). Remember that you can get a maximum of only 10 points for each essay. As you evaluate your work, don't be kind to yourself just because you like your own essay. If the checklist mentions an explanation of structure and function and you failed to give both, then do not give yourself a point. The testing board is very particular about this. You need to mention precisely the things they've listed, in the way they've listed them, in order to gain points.

What if something you've mentioned doesn't appear on the list? Provided you know it's valid, give yourself a point. If it was something you pulled out of your hat at the last minute, odds are it's not directly applicable to the question. However, because you might come up with details even more specific than those contained in the checklist, it's not unlikely that the example or structure you've cited is perfectly valid. If so, go ahead and give yourself the point. Remember that the testing board hires college professors and high school teachers to read your essays, so they'll undoubtedly recognize any legitimate information you slip into the essay.

The second part of the answer key involves short paragraphs. These are not templates; the testing board does not expect you to write this way. They are simply additional tools to help illustrate the things you need to squeeze into your essay in order to rack up the points. They explain in some detail how the various parts of the checklist relate to one another and may give you an idea about how best to integrate them into your own essays come test time.

If you find that you didn't do too well on the essay portion, be sure to carefully read Chapter 14 on the free-response questions. You can use the chapters themselves as checklists. If you find you can't be objective when you grade your own essays, see if your teacher or a classmate will help you out. Good luck!

Long-Form Free-Response Checklist 1

a. **Role of sodium potassium pump**—4 points maximum

> Active transport
> 3 sodium out of the cell
> 2 potassium into the cell
> Potassium leak channels
> Negative inside the cell
> –70 millivolts
> Created gradients

b. **Ion flow**—4 points maximum

> Resting stage (voltage charge is –70 millivolts)
> Trigger stimulus to reach threshold
> Depolarization (Na^+ moves into the cell, voltage-gated channels)
> Repolarization (K^+ ions move out of the cell, voltage-gated channels)
> Back at resting stage

c. **Predicted Outcome**—1 point maximum

Unable to depolarize when threhold is reached

d. **Myelin**—1 point maximum

Faster, because the action potential can jump along the neuron

Sample Long-Form Free Response 1

The sodium potassium pump is responsible for maintaining the resting membrane potential. It actively pumps three Na$^+$ ions out of the cell for every two K$^+$ ions brought into the cell. Since this is active transport, these ions are pumped against the concentration gradient. The outside of the cell has more sodium, and the inside has more potassium. The inside is more negative, since more positives leave the cell than enter the cell. The potassium can also leak out a bit by potassium leak channels.

When a nerve cell is undisturbed, the membrane is said to be in a resting stage. The membrane is polarized, and the voltage charge is −70 millivolts. During a nerve impulse, there is a change in the membrane permeability, and the voltage charge becomes slightly positive. At a certain voltage, voltage-gated sodium channels are opened. Na$^+$ ions rush into the cell, and the inside becomes more positively charged. The membrane is now said to be depolarized. Na$^+$ move into the cell down its concentration gradient. Next, the Na$^+$ channels close, voltage-gated K$^+$ channels open, and K$^+$ ions move out of the cell. The cell is now said to be repolarized as it gets more negative. Then, the original ion concentration is reestablished by the sodium-potassium pump.

If there were not any voltage-gated sodium channels, the action potential couldn't "fire" when threshold is reached, and the cell would not have a rapid depolarization event.

Myelin is like an insulator that covers the outside of some axons. It allows action potentials to move more quickly because the wave of positive charge can jump along the axon.

Long-Form Free-Response Checklist 2

a. **Type of Mutation**—3 points maximum

(1 point each)
Base substitution (involves one nucleotide being replaced by another)
Change in DNA/codon of mRNA causes the wrong tRNA to be recruited
The wrong amino acid gets added to the polypeptide
Change in polypeptide, in turn, alters hemoglobin protein
Distorted red blood cell cannot efficiently carry oxygen

b. **Gel Electrophoresis**—5 points maximum

How it works—3 points maximum
(1 point each)

Apparatus	DNA is put into wells of an agarose gel with a buffer
Electricity	Electrical potential (electrical charge moves fragments)
Charge	Negatively charged fragments move toward the positive pole
Size/Molecular weight	Smaller fragments move faster across the gel

How it can be applied to identify the mutant gene—2 points maximum
(1 point each)
Compare two DNA samples with normal and mutant hemoglobin genes.
Use restriction enzymes to cut the DNA into several fragments.

The normal and the mutant hemoglobin will be cut into different fragments.

Normal hemoglobin can be used as a marker.

When the two samples are run on a gel, they will separate into different banding patterns because they will have been cut differently by the enzyme.

c. **Genotype and Phenotype**—2 points maximum

Sickle-cell disease is an autosomal recessive disorder. Gender does not matter.

A normal female could be heterozygous or homozygous dominant.

A male with the disease would be homozygous.

Sample Long-Form Free Response 2

Sickle-cell anemia is a disease in which the red blood cells have an abnormal shape because of a base substitution. Base substitution involves an error in DNA replication in which one nucleotide is replaced by another nucleotide. This causes the codon of an mRNA to contain an incorrect base. The codon, therefore, matches up with the anticodon of a different tRNA. This tRNA carries a different amino acid. The change in the amino acid alters the polypeptide. The polypeptide, in turn, alters the hemoglobin protein. In this case, the distorted red blood cell cannot efficiently carry oxygen.

Biologists must have determined the nature of the hemoglobin mutation by comparing the normal hemoglobin gene to the abnormal hemoglobin gene, using the technique gel electrophoresis. Gel electrophoresis identifies the difference between two molecules by examining the different rates each molecule moves across the gel. Substances move across a gel according to their molecular weight. For example, smaller fragments move faster than larger fragments. The biologists must have placed fragments of the two genes on an agarose gel and used the normal hemoglobin gene as the marker—the source of comparison to the abnormal hemoglobin gene. (A restriction enzyme was used to cut the two DNA sequences into several fragments prior to loading the gel.) The two DNA sequences should have been identical except for the fragment that contained the abnormal gene.

If a normal noncarrier female mates with a male who is homozygous for the disease, these are the results using a Punnett square.

	H	H
h	Hh	Hh
h	Hh	Hh

All of the offspring would be heterozygotes (genotypes). For sickle-cell anemia, these offspring would be carriers and not actually have the disease phenotype.

Short-Form Free-Response Checklist 3

a. **Cell size ratio**—1 point maximum

    (1 point each)

    A higher ratio of surface area-to-volume allows for greater space for solutes to move in and out of cells.

    Cells must maintain homeostasis and, in order to do this, must eliminate wastes, ingest nutrients, and maintain osmotic and ion balances.

    As a cell grows larger, this ratio diminishes, limiting cell size.

b. **Adaptations**—4 points maximum

    (1 point each)

    Alveoli

    Convoluted membranes in chloroplasts and mitochondria

    Root hairs

    Villi and microvilli

    Endoplasmic reticulum

c. **Transport across the membrane**—1 point maximum

    Small polar molecules cross by facilitated diffusion and require membrane channels.

    Small nonpolar molecules can freely diffuse across the lipid bilayer.

Sample Short-Form Free Response 3

Cells need a certain surface area-to-volume ratio to exchange materials with the environment in order to maintain homeostasis. A high surface area-to-volume ratio allows cells to regulate ion concentrations as well as take in nutrients and eliminate wastes. As a cell becomes larger, the ratio of surface area to volume decreases, and if a cell grows too large, it cannot carry out these functions and therefore will not survive. Organisms have adapted a variety of ways to increase surface area. One example is alveoli, small air sacs in the lungs that maximize surface area in order to allow for maximum respiration. Another example of increasing surface area is villi in the small intestine, whose expansion results in a large surface area for the absorption of nutrients.

Depending on polarity, molecules cross the cell membrane in different ways. Small nonpolar molecules, like carbon dioxide, can freely diffuse through the cell membrane. However, due to the hydrophobic nature of the interior of the cell membrane, small polar molecules need protein channels to allow selective diffusion into or out of the cell. One example of a polar molecule is water, which crosses the membrane through channels called aquaporins.

Short-Form Free-Response Checklist 4

a. Krebs cycle and the electron transport chain, and chemiosmosis as aerobic processes—2 points maximum

> They require oxygen.
> They cannot occur under anaerobic conditions.

b. Site of each step—3 points maximum

Stage	Site
Krebs cycle	Mitochondrial matrix
Electron transport chain	Along the inner mitochondrial membrane
Chemiosmosis	As hydrogens move from the intermembrane space to the mitochondrial matrix

Sample Short-Form Free Response 4

a. The Krebs cycle, electron transport chain, and chemiosmosis are all part of aerobic respiration. During aerobic respiration, glucose is completely converted to CO_2, ATP, and water. These steps are considered aerobic processes because they cannot occur under anaerobic conditions; they require oxygen. Glycolysis, on the other hand, can occur under both aerobic and anaerobic conditions.

b. These three stages of aerobic respiration occur in different parts of the mitochondria. The Krebs cycle occurs in the mitochondrial matrix. Two acetyl-CoA enter the Krebs cycle and produce NADH, $FADH_2$, ATP, and CO_2. The products of the Krebs cycle (NADH and $FADH_2$) are sent to the electron transport chain. The electron transport chain occurs along the inner mitochondrial membrane. The final stage of aerobic respiration—chemiosmosis—occurs as hydrogens move across from the intermembrane space to the mitochondrial matrix.

Short-Form Free-Response Checklist 5

(1 point each)
> Produces somatic cells versus produces gametes
> One round of division versus two rounds of division
> Meiosis only: recombination between homologous chromosomes

Sample Short-Form Free Response 5

The difference between meiosis and mitosis arises because of the different goals of these cellular processes. First, mitosis is designed to reproduce copies of somatic cells, while meiosis is designed to create gametes with greater variability. Second, in order to create the appropriate level of genetic information, meiosis requires two rounds of division, whereas recreating somatic cells only requires one round. Third, in order to increase genetic variability, recombination can occur as part of meiosis between homologous chromosomes. Such mixing is not part of mitosis since the goal is to simply recreate the original cell.

Short-Form Free-Response Checklist 6

(1 point each)

Definition of homologous structures

Relevant example (many possible)

Compare to analogous structures

Sample Short-Form Free Response 6

Homologous structures are those with similar structures because they arose from the same ancestral source. An example of homologous structures would be the arm of a human and the front leg of a cat. Their functions are different in their current form, but the ancestral origin is shared. They are different from analogous structures because they occur in species that have have a common ancestor. Analogous structures evolved separately in two lineages and don't come from a common ancestor.

Short-Form Free-Response Checklist 7

(1 point each)

Transduction

Transformation

(2 points each)

Conjugation

Sample Short-Form Free Response 7

There are three ways in which bacteria transmit genetic information laterally to introduce new phenotypes. These are transduction, transformation, and conjugation. Transduction is the transmission of genetic material from one bacteria to another via a lysogenic virus. Transformation is the uptake of naked DNA from the environment. Conjugation is the cell-to-cell transfer of genetic material in the form of plasmids across a pilus formed between two cells.

Short-Form Free-Response Checklist 8

(1 point each)

No respiration

Not able to create own energy

Not able to reproduce independently

Not able to live outside host cells

Sample Short-Form Free Response 8

Viruses are typically not considered to be alive because they are not able to engage in many of the functions associated with life. Though viruses must engage in processes like translation that require energy, they do not engage in any process to create that energy. Along that same line, there is no respiration of any kind being done by viruses. Further, viruses have no active existence outside of their host cells; no viral functions happen outside of the context of a cell. Due to that restriction, viruses are also not able to reproduce themselves. Being entirely reliant on a host cellular environment for reproduction is another reason to consider them as not alive.

Part III
About the AP
Biology Exam

- The Structure of the AP Biology Exam
- How the AP Biology Exam Is Scored
- Overview of Content Topics
- How AP Exams Are Used
- Other Resources
- Designing Your Study Plan

THE STRUCTURE OF THE AP BIOLOGY EXAM

Beginning in 2018, students may use a four-function, scientific, or graphing calculator.

The AP Biology Exam is three hours long and is divided into two sections: Section I (multiple-choice questions and grid-in quantitative question) and Section II (free-response questions).

Section I consists of 69 questions. These are broken down into Part A (63 multiple-choice questions) and Part B (6 grid-in questions). You will have 90 minutes to complete this section.

Section II involves free-response questions. You'll be presented with two long-form free-response questions and six short-form free-response questions touching upon key issues in biology. You'll be given a 10-minute reading period followed by 80 minutes to answer all eight questions.

If you're thinking that this sounds like a heap of work to try to finish in three hours, you're absolutely right. How can you possibly tackle so much science in so little time? Fortunately, there's absolutely no need to. As you'll soon see, we're going to ask you to leave a small chunk of the test blank. Which part? The parts you don't like. This selective approach to the test, which we call "pacing," is probably the most important part of our overall strategy. But before we talk strategy, let's look at the topics that are covered by the AP Biology Exam.

HOW THE AP BIOLOGY EXAM IS SCORED

AP scores are calculated from your scores on the multiple-choice and free-response sections. The final score is reported on a scale from 1 to 5. The following table explains what that final score means:

Score (Meaning)	Number of test takers receiving this score*	Percentage of test takers receiving this score*	Equivalent grade in a first-year college course	Credit granted for this score?
5 (extremely qualified)	15,661	6.6%	A	Most schools grant credit.
4 (well qualified)	49,944	21%	A–, B+	Most schools grant credit.
3 (qualified)	79,947	33.6%	B–, C	Some schools grant credit; some do not.
2 (possibly qualified)	68,494	28.8%	C, D	Very few schools grant credit.
1 (no recommendation)	24,034	10.1%	D	No schools grant credit.

*The data above is from the College Board website and based on the May 2016 test administration.

Remember that colleges' rules may vary when it comes to granting credit for AP courses. You should contact the individual admissions departments to find out what score you need on the exam to ensure you'll be given credit.

For the AP Biology Exam, your scores on the multiple-choice section and free-response section are each worth 50 percent of your final score. On the multiple-choice section, your total score is based on the number of questions answered correctly, and you do not lose any points for incorrect answers. Unanswered questions do not receive points. The free-response section is graded on a separate point system: long-form responses (Questions 1 and 2) are graded on a 10-point scale, while half of your short-form responses are graded on a 3-point scale and half are scored on a 4-point scale. Your scores are tallied to determine your total free-response score. The free-response score is then combined with your multiple-choice score and weighted to figure out where your score falls within the standard AP scoring scale (1 to 5).

OVERVIEW OF CONTENT TOPICS

The AP Biology Exam covers these four Big Ideas.

These are the *official* Big Ideas:

- **Big Idea 1:** The process of evolution drives the diversity and unity of life.
- **Big Idea 2:** Biological systems utilize free energy and molecular building blocks to grow, to reproduce, and to maintain dynamic homeostasis.
- **Big Idea 3:** Living systems store, retrieve, transmit, and respond to information essential to life processes.
- **Big Idea 4:** Biological systems interact, and these systems and their interactions possess complex properties.

These are the big ideas broken down into simple concepts. As you read this book, think about how these themes fit with various areas of biology:

- Building blocks/hierarchy
- Responses to stimuli
- Organization
- Structure/function
- Communication
- Relationships
- Disruptions and consequences
- Critical thinking

To fully understand the four big ideas, a solid grasp of the following topics is required. These topics include the following:

- Chemistry of Life
  - Important properties of water
  - pH
  - Carbohydrates
  - Proteins
  - Lipids
  - Nucleic acids
  - Origins of life

- Cells
  - Prokaryotic and eukaryotic cells
  - Organelles
  - Membranes and transport
  - Cell junctions
  - Cell communication

- Cellular Energetics
  - Change in free energy
  - Enzymes
  - Coupled reactions and ATP
  - Photosynthesis
  - Cellular respiration (glycolysis, Krebs, oxidative phosphorylation)
  - Fermentation

- Cell Reproduction
  - Cell cycle
  - Mitosis
  - Meiosis

- Molecular Biology
  - DNA and genome structure
  - Transcription
  - Translation
  - Gene regulation
  - Mutation
  - Biotechnology

- Heredity
  - Mendelian genetics
  - Inheritance patterns

- Evolutionary Biology
  - Natural selection
  - Evidence of evolution
  - Phylogenetic trees
  - Impact of genetic variation
  - Speciation
  - Hardy-Weinberg equilibrium

- Structure and Function of Living Things
  - Embryo development and body plan
  - Immune system
  - Viruses and bacteria
  - Nervous system
  - Endocrine system

- Ecology
  - Behavior and communication
  - Food webs and energy pyramids
  - Succession
  - Communities and ecosystems
  - Global issues

- Quantitative Skills
  - Types of data
  - Use of mathematical formulas
  - Descriptive statistics
  - Graphing
  - Hypothesis testing

This might seem like an awful lot of information, but for each topic, there are just a few key facts you'll need to know. Your biology textbooks may go into far greater detail about some of these topics than we do. That's because they're trying to teach you "correct science," whereas we're aiming to improve your scores. Our science is perfectly sound; it's just cut down to size. We've focused on crucial details and given you only what's important. Moreover, as you'll soon see, our treatment of these topics is far easier to handle.

The AP Biology exam not only tests your content knowledge, but it also tests how you apply that knowledge during scientific inquiry. Simply put, the test's authors are testing if you can design and/or think critically about experiments and the hypotheses, evidence, math, data, conclusions, and theories therein. There are seven broad science practices that are tested.

- **Science Practice 1:** The student can use representations and models to communicate scientific phenomena and solve scientific problems.
- **Science Practice 2:** The student can use mathematics appropriately.
- **Science Practice 3:** The student can engage in scientific questioning to extend thinking or to guide investigations within the context of the AP course.
- **Scientific Practice 4:** The student can plan and implement data collection strategies appropriate to a particular scientific question.
- **Scientific Practice 5:** The student can perform data analysis and evaluation of evidence.
- **Science Practice 6:** The student can work with scientific explanations and theories.
- **Science Practice 7:** The student is able to connect and relate knowledge across various scales, concepts, and representations in and across domains.

HOW AP EXAMS ARE USED

Different colleges use AP Exams in different ways, so it is important that you go to a particular college's website to determine how it uses AP Exams. The three items below represent the main ways in which AP Exam scores can be used:

- **College Credit.** Some colleges will give you college credit if you score well on an AP Exam. These credits count toward your graduation requirements, meaning that you can take fewer courses while in college. Given the cost of college, this could be quite a benefit, indeed.
- **Satisfy Requirements.** Some colleges will allow you to "place out" of certain requirements if you do well on an AP Exam, even if they do not give you actual college credits. For example, you might not need to take an introductory-level course, or perhaps you might not need to take a class in a certain discipline at all.

- **Admissions Plus.** Even if your AP Exam will not result in college credit or allow you to place out of certain courses, most colleges will respect your decision to push yourself by taking an AP Course or an AP Exam outside of a course. A high score on an AP Exam shows mastery of more difficult content than is taught in many high school courses, and colleges may take that into account during the admissions process.

OTHER RESOURCES

There are many resources available to help you improve your score on the AP Biology Exam, not the least of which are your **teachers**. If you are taking an AP class, you may be able to get extra attention from your teacher, such as obtaining feedback on your essays. If you are not in an AP course, reach out to a teacher who teaches AP Biology and ask if the teacher will review your essays or otherwise help you with content.

Want to know which colleges are best for you? Check out The Princeton Review's College Advisor app to build your ideal college list and find your perfect college fit! Available for free in the iOS App Store and Google Play Store.

Another wonderful resource is **AP Central**, the official site of the AP Exams. The scope of the information at this site is quite broad and includes:

- a course description, which includes details on what content is covered and sample questions
- sample questions from the AP Biology Exam
- free-response question prompts and multiple-choice questions from previous years

The AP Students home page address is: apstudent.collegeboard.org/home.

For up-to-date information about the ongoing changes to the AP Biology Exam Course, please visit: apstudent.collegeboard.org/apcourse/ap-biology.

Finally, The Princeton Review offers tutoring and small group instruction. Our expert instructors can help you refine your strategic approach and add to your content knowledge. For more information, call 1-800-2REVIEW.

DESIGNING YOUR STUDY PLAN

In Part I, you identified some areas of potential improvement. Let's now delve further into your performance on Practice Test 1 with the goal of developing a study plan appropriate to your needs and time commitment.

Read the answers and explanations associated with the multiple-choice questions (starting at page 33). After you have done so, respond to the following questions:

- Review the bulleted list of topics on pages 50 and 51. Next to each topic, indicate your rank of the topic as follows: 1 means "I need a lot of work on this," 2 means "I need to beef up my knowledge," and 3 means "I know this topic well."

- How many days/weeks/months away is your exam?

- What time of day is your best, most focused study time?

- How much time per day/week/month will you devote to preparing for your exam?

- When will you do this preparation? (Be as specific as possible: Mondays and Wednesdays from 3:00 to 4:00 P.M., for example.)

- Based on the answers above, will you focus on strategy (Part IV), content (Part V), or both?

- What are your overall goals in using this book?

Use those answers to create a study plan. Start with the topics that need the most work and map out when you will study and what you will study.

Remember, your schedule may evolve along the way. If a certain time/location is not working for you, then try mixing it up. If you are struggling with a topic, perhaps try tackling it with a teacher, tutor, or a classmate.

Part IV
Test-Taking Strategies for the AP Biology Exam

PREVIEW

Review your responses to the three questions on page 2 and then respond to the following questions:

- How many multiple-choice questions did you miss even though you knew the answer?

- On how many multiple-choice questions did you guess blindly?

- How many multiple-choice questions did you miss after eliminating some answers and guessing based on the remaining answers?

- Did you find any of the free-response questions easier/harder than the others—and, if so, why?

HOW TO USE THE CHAPTERS IN THIS PART

For the following Strategy chapters, think about what you are doing now before you read the chapters. As you read and engage in the directed practice, be sure to appreciate the ways you can change your approach. At the end of Part IV, you will have the opportunity to reflect on how you will change your approach.

Chapter 1
How to Approach
Multiple-Choice
Questions

SECTION I

As we mentioned earlier, the multiple-choice section consists of the following two parts:

- Part A—63 multiple-choice questions dealing with an experiment or a set of data or with general knowledge of biology
- Part B—6 grid-in questions

Part A

Part A of Section I consists of 63 run-of-the-mill multiple-choice questions. These questions test your grasp of the fundamentals of biology and your ability to apply biological concepts to help solve problems. Here's an example:

> If a coding segment of DNA reads 5'-ATG-CCA-GCT-3', the mRNA strand that results from the transcription of this segment will be
>
> (A)   3'-TAC-GGT-CGA-5'
> (B)   3'-UCG-ACC-GUA-5'
> (C)   3'-UAA-GGU-CGA-5'
> (D)   3'-TAC-GGT-CTA-5'

Don't worry about the answer to this question for now. By the end of this book, it will be a piece of cake. The majority of the questions in this section are presented in this format. A few questions may include a figure, a diagram, or a chart.

The second part of the first portion also consists of multiple-choice questions, but here you're asked to think logically about different biological experiments or data. Here's a typical example:

Questions 60 and 61 refer to the following diagram and information.

To understand the workings of neurons, an experiment was conducted to study the neural pathway of a reflex arc in frogs. A diagram of a reflex arc is given below.

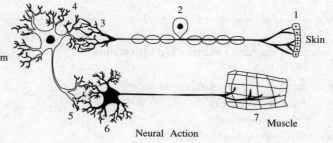

60. Which of the following represents the correct pathway taken by a nerve impulse as it travels from the spinal cord to effector cells?

(A) 1-2-3-4
(B) 6-5-4-3
(C) 2-3-4-5
(D) 4-5-6-7

61. The brain of the frog is destroyed. A piece of acid-soaked paper is applied to the frog's skin. Every time the piece of paper is placed on its skin, one leg moves upward. Which of the following conclusions is best supported by the experiment?

(A) Reflex actions are not automatic.
(B) Some reflex actions can be inhibited.
(C) All behaviors in frogs are primarily reflex responses.
(D) This reflex action does not require the brain.

You'll notice that these particular questions refer to an experiment. Many of the questions in this portion test your ability to integrate information, interpret data, and draw conclusions from the results.

Part B

Part B of Section I consists of six grid-in questions that require using information presented in the question to calculate an answer. That numeric response is then filled in on a grid and bubbled accordingly. Answers can be in the form of integers, decimals, or fractions. A four-function (with square root), scientific, or graphing calculator may be used on these questions. Here's a typical example:

Popular Topics for Quantitative Questions

- Surface area-to-volume ratio
- Hardy-Weinberg
- Water potential
- Energy pyramids/biomass
- Chi-squared analysis
- Gene linkage
- Inheritance/probability

If the genotype frequencies of an insect population are AA = 0.49, Aa = 0.42, and aa = 0.09, what is the gene frequency of the dominant allele?

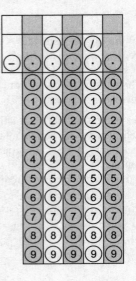

We mentioned earlier that our approach is strategy-based. As you're about to see, many of these strategies are based on common sense—for example, using mnemonics like "ROY G. BIV." (Remember that one? It's the mnemonic for red, orange, yellow, green, blue, indigo, violet—the colors of the spectrum of visible light.) Other strategies do not make as much sense. In fact, we're going to ask you to throw out much of what you've been taught when it comes to taking standardized tests.

PACE YOURSELF

When you take a test in school, how many questions do you answer? Naturally, you try to answer all of them. You do this for two reasons: (1) your teacher told you to, and (2) if you left a question blank, your teacher would mark it wrong. However, that's not the case when it comes to the AP Biology Exam. In fact, finishing the test is the worst thing you can do. Before we explain why, let's talk about timing.

One of the main reasons that taking the AP Biology Exam can be so stressful is the time constraint we discussed above—75 seconds per multiple-choice question, 20 minutes per long essay, and 6 minutes per short essay. If you had all day, you would probably ace the test. We can't give you all day, but we can do the next best thing: we can give you more time for each question. How? By having you slow down and answer fewer questions.

Slowing down and doing well on the questions you do answer is the best way to improve your score on the AP Biology Exam. Rushing through questions in order to finish, on the other hand, will always hurt your score. When you rush, you're far more likely to make careless errors, misread, and fall into traps. Keep in mind that blank answers are not counted against you.

3 is the Magic Number
Using the Three-Pass System will help you earn more points by doing the questions you know first and saving the ones you don't know for last.

THE THREE-PASS SYSTEM

The AP Biology Exam covers a broad range of topics. There's no way, even with our extensive review, that you will know everything about every topic in biology. So what should you do?

Do the Easiest Questions First

The best way to rack up points is to focus on the questions you find the easiest. If you know the answer, nail it and move on. Other questions, however, will be a little more complicated. As you read each question, decide if you think it's easy, medium, or hard. During a first pass, do all of your easy questions. If you come across a problem that seems time-consuming or completely incomprehensible, skip it. Remember:

> Questions that you find easy are worth just as many points as
> the ones that stump you, so your time is better spent focusing
> on the ones in the former group.

Save the medium questions for the second pass. These questions are either time-consuming or require that you analyze all the answer choices (i.e., the correct answer doesn't pop off the page). If you come across a question that makes no sense from the outset, save it for the last pass. You're far less likely to fall into a trap or settle on a silly answer.

Watch Out for Those Bubbles!

Since you're skipping problems, you need to keep careful track of the bubbles on your answer sheet. One way to accomplish this is by answering all the questions on a page and then transferring your choices to the answer sheet. If you prefer to enter them one by one, make sure you double-check the number beside the ovals before filling them in. We'd hate to see you lose points because you forgot to skip a bubble!

So then, what about the questions you don't skip?

The Art of POE

You can earn a great score on the multiple-choice section by applying POE. Eliminating wrong answers can be just as effective as knowing the correct answer right off the bat.

PROCESS OF ELIMINATION (POE)

On most tests, you need to know your material backward and forward to get the right answer. In other words, if you don't know the answer beforehand, you probably won't answer the question correctly. This is particularly true of fill-in-the-blank and essay questions. We're taught to think that the only way to get a question right is by knowing the answer. However, that's not the case on Section I of the AP Biology Exam. You can get a perfect score on this portion of the test without knowing a single right answer, provided you know all the wrong answers!

What are we talking about? This is perhaps the single most important technique in terms of the multiple-choice section of the exam. Let's take a look at the example below.

The structures that act as the sites of gas exchange in a woody stem are the

(A) lungs
(B) lenticels
(C) ganglia
(D) lentil beans

Now if this were a fill-in-the-blank-style question, you might be in a heap of trouble. But let's take a look at what we've got. You see "woody stem" in the question, which leads you to conclude that the question is about plants. Right away, you know the answer is not (A) or (C) because plants don't have lungs or ganglia. Now you've narrowed it down to (B) and (D). Notice that (B) and (D) look very similar. Obviously, one of them is a trap. At this point, if you don't know what "lentil beans" are, you have to guess. However, even if you don't know precisely what they are, it's safe to say that you probably know that lentil beans have nothing to do with plant respiration. Therefore, the correct answer is (B), lenticels.

Although our example is a little goofy and doesn't look exactly like the questions you'll be seeing on the test, it illustrates an important point:

> **Process of Elimination (POE)** is the best way to approach the multiple-choice questions.

Even if you don't know the answer right off the bat, you'll surely know that two or three of the answer choices are not correct. What then?

Aggressive Guessing

The testing board, the service that develops and administers the exam, tells you that random guessing will not affect your score. This is true. There is no guessing penalty on the AP Biology Exam. For each correct answer you'll receive one point and you will not lose any points for each incorrect answer.

Although you won't lose any points for wrong answers, you should guess aggressively by getting rid of the incorrect answer choices. The moment you've eliminated a couple of answer choices, your odds of getting the question right, even if you guess, are far greater. If you can eliminate as many as two answer choices, your odds improve enough that it's in your best interest to guess.

WORD ASSOCIATIONS

Another way to rack up the points on the AP Biology Exam is by using word associations in tandem with your POE skills. Make sure that you memorize the words in the Key Terms lists throughout this book. Know them backward and forward. As you learn them, make sure you group them by association, since the testing board is bound to ask about them on the AP Biology Exam. What do we mean by "word associations"? Let's take the example of mitosis and meiosis.

You'll soon see from our review that there are several terms associated with mitosis and meiosis. *Synapsis*, *crossing-over*, and *tetrads*, for example, are words associated with meiosis but not mitosis. We'll explain what these words mean later in this book. For now, just take a look.

Which of the following typifies cytokinesis during mitosis?

(A)   Crossing-over
(B)   Formation of tetrads
(C)   Synapsis
(D)   Division of the cytoplasm

This might seem like a difficult problem. But let's think about the associations we just discussed. The question asks us about mitosis. However, (A), (B), and (C) all mention events that we've associated with meiosis. Therefore, they are out. Without even racking your brain, you've managed to find the correct answer: (D). Not bad!

Once again, don't worry about the science for now. We'll review it later. What is important to recognize is that by combining the associations we'll offer throughout this book and your aggressive POE techniques, you'll be able to rack up points on problems that might have seemed difficult at first.

MNEMONICS—OR THE BIOLOGY NAME GAME

One of the big keys to simplifying biology is the organization of terms into a handful of easily remembered packages. The best way to accomplish this is by using mnemonics. Biology is all about names: the names of chemical structures, processes, theories, and so on. How are you going to keep them all straight? A mnemonic, as you may already know, is a convenient device for remembering something.

For example, one important issue in biology is taxonomy, that is, the classification of life-forms, or organisms. Organisms are classified in a descending system of similarity, leading from domains (the broadest level) to species (the most specific level). The complete order runs as follows: domain, kingdom, phylum, class, order, family, genus, and species. Don't freak out yet. Use a mnemonic to help you:

King Philip of Germany decided to walk to America. What do you think happened?

Dumb	→	Domain
King	→	Kingdom
Philip	→	Phylum
Came	→	Class
Over	→	Order
From	→	Family
Germany	→	Genus
Soaked	→	Species

Learn the mnemonic and you'll never forget the science!

Mnemonics can be as goofy as you like, as long as they help you remember. Throughout this book, we'll give you mnemonics for many of the complicated terms we'll be seeing. Use ours, if you like them, or feel free to invent your own. Be creative! Remember: the important thing is that you remember the information, not how you remember it.

IDENTIFYING EXCEPT QUESTIONS

About 10 percent of the multiple-choice questions in Section I are EXCEPT/ NOT/LEAST questions. With this type of question, you must remember that you're looking for the *wrong* (or the least correct) answer. The best way to go about these is by using POE.

More often than not, the correct answer is a true statement, but is wrong in the context of the question. However, the other three tend to be pretty straightforward. Cross off the three that apply, and you're left with the one that does not. Here's a sample question.

> All of the following are true statements about gametes EXCEPT
>
> (A) they are haploid cells
> (B) they are produced only in the reproductive structures
> (C) they bring about genetic variation among offspring
> (D) they develop from polar bodies

Outsmart the Tricky Questions
EXCEPT and NOT questions are considered more difficult than other question types. Here's where POE comes in handy!

If you don't remember anything about gametes and gametogenesis, or the production of gametes, this might be a particularly difficult problem. We'll see these again later on, but for now, remember that gametes are the sex cells of sexually reproducing organisms. As such, we know that they are haploid and are produced in the sexual organs. We also know that they come together to create offspring.

From this very basic review, we know immediately that (A) and (B) are not our answers. Both of these are accurate statements, so we eliminate them. That leaves us with (C) and (D). If you have no idea what (D) means, focus on (C). In sexual reproduction, each parent contributes one gamete, or half the genetic complement of the offspring. This definitely helps vary the genetic makeup of the offspring. Choice (C) is a true statement, so it can be eliminated. The correct answer is (D).

Don't sweat it if you don't recall the biology. We'll be reviewing it in detail soon enough. For now, remember that the best way to answer these types of questions is to spot all the right statements and cross them off. You'll wind up with the wrong statement, which happens to be the correct answer.

SOLVING GRID-IN QUESTIONS

Part B of the multiple-choice section requires you to answer six grid-in questions. Unlike the questions in Part A, you will not be able to use Process of Elimination. However, there are strategies you can use to increase your chances of efficiently getting these questions correct. First, identify what the question is actually asking for. Are you being asked to find a value on a graph or chart? Perhaps they are having you calculate the frequency of an allele using Hardy-Weinberg equilibrium. If you have no idea how to find out what they are asking you, take a quick educated guess and move on. For instance, if they are asking for a frequency, try guessing a decimal between 0 and 1, such as 0.25. However, if you do know how to find the answer, then write down the question number and any work (notes, calculations, etc.) that you would need to answer the question neatly. Before filling in your answer, make sure that you haven't made any mistakes on interpreting data or making calculations. Be sure to bubble in your answers.

Remember Your Resources!

The AP Biology Equations and Formulas sheet is included at the end of this book and in your online Student Tools. Print it out and remember to use it when you take the Practice Tests.

Although these grid-in questions require calculations, fear not! The College Board provides you with a handy formula sheet, which you can find at the end of this book. You can expect to be given this sheet as part of the exam, so you do not need to memorize these equations. However, it will be important for you to practice using them and familiarizing yourself with the formula sheet prior to the exam. Keep this page handy as you review and practice problems so that you will be accustomed to using it on the day of the exam.

Chapter 2
How to Approach
Free-Response
Questions

THE ART OF THE FREE-RESPONSE ESSAY

On Section II of the exam, you are given two long-form free-response questions and six short-form free-response questions to answer in 80 minutes. You will also have a 10-minute reading period, which means you have a total of 90 minutes to complete Section II. The best way to rack up points on this section is to give the readers what they're looking for. Fortunately, we know precisely what that is.

The readers, or graders, have a checklist of key terms and concepts that they use to assign points. We like to call these "hot button" terms. Simply put, for each hot button term that you include in a response, you will receive a predetermined number of points. For example, if the question deals with the function of enzymes, the graders are instructed to give two points for a mention of the "lock-and-key theory of enzyme specificity."

Naturally, you can't just compose a "laundry list" of scientific terms. Otherwise, it wouldn't be an essay. What you can do, however, is organize your free-response essays around a handful of these key points. The most effective and efficient way to do this is by using the 10-minute reading period to brainstorm and come up with the scientific terms. Then outline your response before you begin to write, using your hot buttons as your guide.

READ THE QUESTIONS CAREFULLY

Make the Most of Your Planning Time
If you use these 10 minutes wisely, you can breeze your way through the free-response questions.

The testing board gives you 10 minutes to read the questions and organize your thoughts before you begin writing. The first thing you should do is take less than a minute to skim all of the questions and put them into your own personal order of difficulty from easiest to toughest. Once you've decided the order in which you will answer the questions (easiest first, hardest last), you can begin to formulate your responses. Your first step should be a more detailed assessment of each question.

The most important advice we can give you is to read each question at least twice. As you read the question, focus on key words, especially "direction words." Almost every essay question begins with a direction word. These words are often in bold, so they should jump out at you. Some examples of direction words are *discuss*, *define*, *explain*, *describe*, *compare*, and *contrast*. If a question asks you to discuss a particular topic, you should give a viewpoint and support it with examples. If a question asks you to compare two things, you should discuss how the two things are similar. On the other hand, if the question asks you to contrast two things, you need to show how these things are different.

Many students lose points on their essays because they either misread the question or fail to do what's asked of them.

BRAINSTORM

Your next objective during the 10-minute reading period should be to organize your thoughts.

Once you've read the questions, you need to brainstorm. Jot down as many key terms and concepts as you can. Don't forget, the test readers assign points on the basis of these key concepts. For each one that you mention and/or explain, you get a point.

How many do you need? You won't need all of them, that's for sure. However, you will need enough to get you the maximum number of points for that question. Most of these can be pulled directly from your reading in this book. At the end of each chapter is a Key Terms list, which is full of terms that you should incorporate into your free responses. In addition to this list, you can feel free to create your own thorough lists of terms for your own use.

Don't spend more than about 2 minutes per question brainstorming. Once your 10-minute reading period is up, you should be ready to start writing your essays.

OUTLINE YOUR RESPONSE

Have you ever written yourself into a corner? You're halfway through your essay when you suddenly realize that you have no idea whatsoever where you're going with your train of thought. To avoid this (and the panic that accompanies it), take a few minutes to draft an outline.

Your outline should incorporate as many of the hot buttons as you need in order to maximize your score. In other words, if a question asks for two examples, choose only those two with which you are most comfortable. In your outline, make notes about the crucial points to mention with regard to each topic or key word. Once your outline is complete, you're ready to move on to writing the actual essay.

All of this preparation may seem time-consuming. However, it should take no more than two minutes per long question and one minute per short question. What's more, it will greatly simplify the whole essay-writing process. So while you lose a little time at the outset, you'll more than make up for it when it comes time to actually write your answer.

DEVELOP YOUR IDEAS IN EACH RESPONSE

Now you can use your outline to write your essay. Most students can come up with key terms or phrases that concern a particular biology topic. What separates a high-scoring student from a low-scoring student is *how* the student develops his or her thoughts in each essay. Besides giving the hot buttons, you'll need to elaborate on your thoughts and ideas. For example, don't just throw out a list of terms that pertain to meiosis and mitosis (such as *synapsis*, *crossing-over*, and *gametes*). Go one step further. Make sure you mention the *significance* of meiosis (it produces genetic variability). This extra piece of information will earn you an extra point.

Generally, you'll need to write about two to four paragraphs, depending on the number of parts contained in each essay question. In addition, be sure to give the appropriate number of examples for each essay. If the question asks for three examples, give only three examples. If you present more than is required, the test readers won't even read them or count them toward your score. The bottom line is this: stick to the question.

ANSWER EACH PART OF THE QUESTION SEPARATELY

The more parts there are to a free-response question, the more important it is to pace yourself. For each question, you're better off writing a little bit for each part than you are spending all your time on any one part of a question. Why? Even if you were to write the perfect answer to one part of a question, there's a limit to the number of points the test readers can assign to that part.

As you address each part of the question, make sure you break your response into paragraphs to make the test reader's job easier. When test readers have an easy time reading your essay, they're more likely to award points: your writing comes across as clearer and more organized. Readers have also requested that students label each part of their essay answers, so write "Part (a)," "Part (b)," and so on, accordingly in your response.

Finally, don't spend too much time writing a fancy introduction. You won't get brownie points for beautifully written openings like, "It was the best of experiments, it was the worst of experiments." Just leap right into the essay. And don't worry too much about grammar or spelling errors. Your grammar can hurt only if it's so bad that it seriously impairs your ability to communicate.

Head over to the College Board
For sample free-response questions and more, check out the AP Students page on the College Board website.

INCORPORATE ELEMENTS OF AN EXPERIMENTAL DESIGN

Since one of the free-response questions will be experimentally based, you'll need to know how to design an experiment. Most of these questions require that you present an appropriately labeled diagram or graph. Otherwise, you'll get only partial credit for your work.

There are two things you must remember when designing experiments on the AP Biology Exam: (1) always label your figures, and (2) include controls in all experiments. Let's take a closer look at these two points.

Know How to Label Diagrams and Figures

Let's briefly discuss the important elements in setting up a graph. The favorite type of graph on the AP Biology Exam is the **coordinate graph**. The coordinate graph has a horizontal axis (*x*-axis) and a vertical axis (*y*-axis).

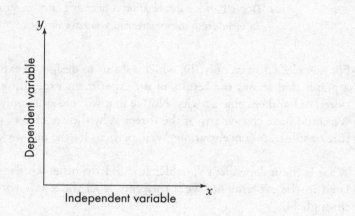

The *x*-axis usually contains the **independent variable**—the thing that's being manipulated or changed. The *y*-axis contains the **dependent variable**—the thing that's affected when the independent variable is changed.

Now let's look at what happens when you put some points on the graph. Every point on the graph represents both an independent variable and a dependent variable.

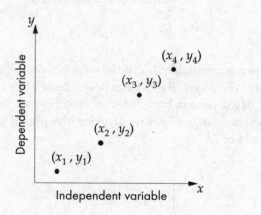

Once you draw both axes and label the axes as *x* and *y*, you can plot the points on the graph. Let's look at the following question.

Enzymes are biological catalysts.

(a) **Relate** the chemical structure of an enzyme to its catalytic activity and specificity.

(b) **Design** an experiment that investigates the influence of temperature, substrate concentration, or pH on the activity of an enzyme.

(c) **Describe** what information concerning enzyme structure could be inferred from the experiment you have designed.

For now, let's discuss only (b), which asks us to design an experiment. Let's set up a graph that shows the results of an experiment examining the relationship between pH and enzyme activity. Notice that we've chosen only one factor here, pH. We could have chosen any of the three. Why did we choose pH and not temperature or substrate concentration? Well, perhaps it's the one we know the most about.

What is the independent variable? It is pH. In other words, pH is being manipulated in the experiment. We'll therefore label the *x*-axis with pH values from 0 through 14.

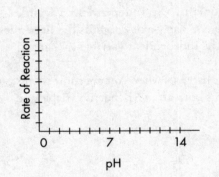

What is the dependent variable? It's the enzyme activity—the thing that's affected by pH. Let's label the *y*-axis "Rate of Reaction." Now we're ready to plot the values on the graph. Based on our knowledge of enzymes, we know that for most enzymes the functional range of pH is narrow, with optimal performance occurring at or around a pH of 7.

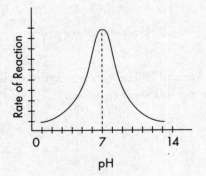

Now you should interpret your graph. If the pH level decreases from a neutral pH of 7, the reaction rate of the enzyme will decrease. If the pH level increases, the rate of reaction will also decrease. Don't forget to include a simple explanation of your graph.

Additional quantitative and graphing skills will be discussed at length in a later chapter of this text. There you will learn how to include statistics and hypothesis testing in your free-response questions and how best to present data using graphs.

INCLUDE CONTROLS IN YOUR EXPERIMENTS

Almost every experiment will have at least one group that represents the normal/unchanged/neutral version of the independent variable. This is called the *control*. A control is simply a standard of comparison. What does a control do? It enables the biologist to verify that the outcome of the study is due to changes in the independent variable and nothing else.

Let's say the principal of your school thinks that students who eat breakfast do better on the AP Biology Exam than those who don't eat breakfast. He gives a group of 10 students from your class free breakfast every day for a year. When the school year is over, he administers the AP Biology Exam and they all score brilliantly! Did they do well because they ate breakfast every day? We don't know for sure. Maybe those kids were the smartest kids in the class and would have scored well anyway.

In this case, the best way to be sure that eating breakfast made a difference is to have a control group. In other words, he would need to also follow for a year a group of students of equal intelligence to the first group, BUT who are known to *never* eat breakfast. At the end of that year, he could have them take the AP Biology Exam. If they do just as well as the group of similar intelligence that ate breakfast, then we can probably conclude that eating breakfast wasn't necessarily leading to higher AP scores. The group of students that didn't eat breakfast is called the control group because those students were not "exposed" to the variable of interest—in this case, breakfast. Control groups are generally any group that you include just for the sake of comparison.

Put It All Together
Use the techniques discussed in this chapter and practice writing your free responses. Additional sample free-response questions are available on the AP Students page for the AP Biology Exam.

Chapter 3
Using Time
Effectively to
Maximize Points

BECOMING A BETTER TEST TAKER

Very few students stop to think about how to improve their test-taking skills. Most assume that if they study hard, they will test well and that if they do not study, they will do poorly. Most students continue to believe this even after experience teaches them otherwise. Have you ever studied really hard for an exam and then blown it on test day? Have you ever aced an exam for which you thought you weren't well prepared? Most students have had one, if not both, of these experiences. The lesson should be clear: factors other than your level of preparation influence your final test score. This chapter will provide you with some insights that will help you perform better on the AP Biology Exam and on other exams as well.

PACING AND TIMING

A big part of scoring well on an exam is working at a consistent pace. The worst mistake made by inexperienced or less savvy test takers is that they come to a question that stumps them and rather than just skip it, they panic and stall. Time seems to stand still when you're working on a question you cannot answer, and it is not unusual for students to waste five minutes on a single question (especially a question involving a graph or the word EXCEPT) because they are too stubborn to cut their losses. It is important to be aware of how much time you have spent on a given question and section. There are several ways to improve your pacing and timing for the test:

- **Know your average pace.** While you prepare for your test, try to gauge how long you take on 5, 10, or 20 questions. Knowing how long you spend on average per question will help you identify how many questions you can answer effectively and how best to pace yourself for the test.

- **Have a watch or clock nearby.** You are permitted to have a watch or clock nearby to help you keep track of time. However, it's important to remember that constantly checking the clock is in itself a waste of time and can be distracting. Devise a plan. Try checking the clock every fifteen or thirty questions to see if you are keeping the correct pace or whether you need to speed up. This will ensure that you're cognizant of the time but will not permit you to fall into the trap of dwelling on it.

- **Know when to move on.** Since all questions are scored equally, investing appreciable amounts of time on a single question is inefficient and can potentially deprive you of the chance to answer easier ones later on. You should eliminate answer choices if you are able to, but don't feel bad about picking a random answer and moving on if you cannot find the correct answer. Remember, tests are like marathons;

you do best when you work through them at a steady pace. You can always come back to a question you don't know. When you do, very often you will find that your previous mental block is gone and you will wonder why the question perplexed you the first time around (as you gleefully move on to the next question). Even if you still don't know the answer, you will not have wasted valuable time you could have spent on easier questions.

- **Be selective.** You don't have to do any of the questions in a given section in order. If you are stumped by an essay or multiple-choice question, skip it or choose a different one. In the next section, you will see that you may not have to answer every question correctly to achieve your desired score. Select the questions or essays that you can answer and work on them first. This will make you more efficient and give you the greatest chance of getting the most questions correct.

- **Use Process of Elimination (POE) on multiple-choice questions.** Many times, one or more answer choices can be eliminated. Every answer choice that can be eliminated increases the odds that you will answer the question correctly.

Remember, when all the questions on a test are of equal value, no one question is that important, and your overall goal for pacing is to get the most questions correct. Finally, you should set a realistic goal for your final score. In the next section, we will break down how to achieve your desired score and how to pace yourself to do so.

GETTING THE SCORE YOU WANT

Depending on the score you need, it may be in your best interest not to try to work through every question. Check with the schools to which you are applying to find out what scores they will accept for credit.

AP Exams in all subjects no longer include a "guessing penalty" of a quarter of a point for every incorrect answer. Instead, students are assessed only on the total number of correct answers. A lot of AP materials, even those you receive in your AP class, may not include this information. It's really important to remember that if you are running out of time, you should fill in all the bubbles before the time for the multiple-choice section is up. Even if you don't plan to spend a lot of time on every question or even if you have no idea what the correct answer is, you need to fill something in.

TEST ANXIETY

Everybody experiences anxiety before and during an exam. To a certain extent, test anxiety can be helpful. Some people find that they perform more quickly and efficiently under stress. If you've ever pulled an all-nighter to write a paper and ended up doing good work, you know the feeling.

However, too much stress is definitely a bad thing. Hyperventilating during the test, for example, almost always leads to a lower score. If you find that you stress out during exams, here are a few preemptive actions you can take:

- **Take a reality check.** Evaluate your situation before the test begins. If you have studied hard, remind yourself that you are well prepared. Remember that many others taking the test are not as well prepared, and (in your classes, at least) you are being graded against them, so you have an advantage. If you didn't study, accept the fact that you will probably not ace the test. Make sure you get to every question you know something about. Don't stress out or fixate on how much you don't know. Your job is to score as high as you can by maximizing the benefits of what you do know. In either scenario, it's best to think of a test as if it were a game. How can you get the most points in the time allotted to you? Always answer questions you can answer easily and quickly before tackling those that will take more time.

- **Try to relax.** Slow, deep breathing works for almost everyone. Close your eyes, take a few slow, deep breaths, and concentrate on nothing but your inhalation and exhalation for a few seconds. This is a basic form of meditation that should help you to clear your mind of stress and, as a result, concentrate better on the test. If you have ever taken yoga classes, you probably know some other good relaxation techniques. Use them when you can (obviously, anything that requires leaving your seat and, say, assuming a handstand position won't be allowed by any but the most free-spirited proctors).

- **Eliminate as many surprises as you can.** Make sure you know where the test will be given, when it starts, what type of questions are going to be asked, and how long the test will take. You don't want to be worrying about any of these things on test day or, even worse, after the test has already begun.

The best way to avoid stress is to study both the test material and the test itself. Congratulations! By reading this book, you are taking a major step toward a stress-free AP Biology Exam.

REFLECT

Respond to the following questions:

- How long will you spend on multiple-choice questions?

- How will you change your approach to multiple-choice questions?

- What is your multiple-choice guessing strategy?

- How much time will you spend on the free-response questions?

- What will you do before you begin writing your free-response answers?

- Will you seek further help outside of this book (such as a teacher, tutor, or AP Students) on how to approach the questions that you will see on the AP Biology Exam?

Part V
Content Review
for the AP
Biology Exam

Chapter 4
Chemistry of Life

ELEMENTS

Although life forms exist in many diverse forms, they all have one thing in common: they are all made up of matter. Matter is made up of elements. **Elements**, by definition, are substances that cannot be broken down into simpler substances by chemical means.

THE ESSENTIAL ELEMENTS OF LIFE

Although there are 92 natural elements, 96% of the mass of all living things is made up of just four of them: **oxygen** (O), **carbon** (C), **hydrogen** (H), and **nitrogen** (N). Other elements such as calcium (Ca), phosphorus (P), potassium (K), sulfur (S), sodium (Na), Chlorine (Cl), and magnesium (Mg) are also present, but in smaller quantities. These elements make up most of the remaining four percent of a living thing's weight. Some elements are known as **trace elements** because they are required by an organism only in very small quantities. Trace elements include iron (Fe), iodine (I), and copper (Cu).

SUBATOMIC PARTICLES

The smallest unit of an element that retains its characteristic properties is an **atom**. Atoms are the building blocks of the physical world.

Within atoms, there are even smaller subatomic particles called **protons**, **neutrons**, and **electrons**. Let's take a look at a typical atom.

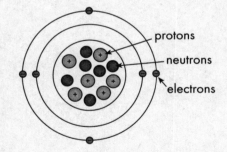

Protons and neutrons are particles that are packed together in the core of an atom called the **nucleus**. You'll notice that protons are positively charged (+) particles, whereas neutrons are uncharged particles.

Electrons, on the other hand, are negatively charged (–) particles that spin around the nucleus. Electrons are pretty small compared to protons and neutrons. In fact, for our purposes, electrons are considered massless. Most atoms have the same number of protons and electrons, making them electrically neutral. Some atoms have the same number of protons but differ in the number of neutrons in the nucleus. These atoms are called **isotopes**.

Carbon-14 Isotopes
Some isotopes are radioactive and decay predictably over time. Ancient artifacts can be dated by examining the rate of decay of carbon-14 and other isotopes within the artifact. This process is called **radiometric dating**.

COMPOUNDS

When two or more individual elements are combined in a fixed ratio, they form a chemical **compound**. You'll sometimes find that a compound has different properties from those of its elements. For instance, hydrogen and oxygen exist in nature as gases. Yet when they combine to make water, they often pass into a liquid state. When hydrogen atoms get together with oxygen atoms to form water, we've got a **chemical reaction**.

$$2H_2\ (g) + O_2\ (g) \rightarrow 2H_2O\ (l)$$

The atoms of a compound are held together by **chemical bonds**, which may be **ionic bonds**, **covalent bonds**, or **hydrogen bonds**.

An ionic bond is formed between two atoms when one or more electrons are transferred from one atom to the other. In order for this to occur, first, one atom *loses* electrons and becomes positively charged, and the other atom *gains* electrons and becomes negatively charged. The charged forms of the atoms are called **ions**. An ionic bond results from the attraction between the two oppositely charged ions. For example, when Na reacts with Cl, the charged ions Na^+ and Cl^- are formed.

A covalent bond is formed when electrons are *shared* between atoms. If the electrons are shared equally between the atoms, the bond is called **nonpolar covalent**. If the electrons are shared unequally, the bond is called **polar covalent**. When one pair of electrons is shared between two atoms, the result is a single covalent bond. When two pairs of electrons are shared, the result is a double covalent bond. When three pairs of electrons are shared, the result is a triple covalent bond.

WATER: THE VERSATILE MOLECULE

One of the most important substances in nature is water. Water is considered a unique molecule because it plays an important role in chemical reactions.

Let's take a look at one of the properties of water. Water has two hydrogen atoms joined to an oxygen atom.

In water molecules, the electrons are not shared equally in the bonds between hydrogen and oxygen. This means that the hydrogen atoms have a partial positive charge and the oxygen atom has a partial negative charge. Molecules that have partially positive and partially negative charges are said to be **polar**. Water is therefore a polar molecule. The positively charged elements of the water molecules strongly attract the negatively charged ends of other polar compounds (including water). Likewise, the negatively charged ends strongly attract the positively charged ends of polar compounds. These forces are most readily apparent in the tendency of water molecules to stick together, as in the formation of water beads or raindrops.

Water Weight
Did you know that more than 60 percent of your body weight consists of water?

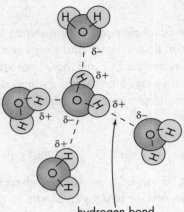

hydrogen bond

These types of intermolecular attractions are called hydrogen bonds. Hydrogen bonds are weak chemical bonds that form when a hydrogen atom that is covalently bonded to one electronegative atom is also attracted to another electronegative atom. Water molecules are held together by hydrogen bonds.

Although hydrogen bonds are individually weak, they are strong when present in large numbers. Because water reacts well with other polar substances, it makes a great solvent; it can dissolve many kinds of substances. The hydrogen bonds that hold water molecules together contribute to a number of special properties, including **cohesion**, **adhesion**, **surface tension**, **high heat capacity**, and **expansion on freezing**.

Cohesion and Adhesion

As mentioned above, water molecules have a strong tendency to stick together. That is, water exhibits cohesive forces. These forces are extremely important to life. For instance, during transpiration, water molecules evaporate from a leaf, "pulling" on neighboring water molecules. These, in turn, draw up the molecules immediately behind them, and so on, all the way down the plant vessels. The resulting chain of water molecules enables water to move up the stem.

Water molecules also like to stick to other substances—that is, they're **adhesive**. Have you ever tried to separate two glass slides stuck together by a film of water? They're difficult to separate because of the water sticking to the glass surfaces.

These two forces taken together—cohesion and adhesion—account for the ability of water to rise up the roots, trunks, and branches of trees. Since this phenomenon occurs in thin vessels, it's called **capillary action**.

Surface Tension

The cohesion of water molecules contributes to another property of water, its **surface tension**. Like a taut trampoline, the surface of water has a tension to it. The water molecules are stuck together and light things like leaves and water striders can sit atop the surface without sinking.

High Heat Capacity

Another remarkable property of water is its high heat capacity. What's heat capacity? Your textbook will give you a definition something like this: "heat capacity is the quantity of heat required to change the temperature of a substance by 1 degree." What does that mean? In plain English, heat capacity refers to the ability of a substance to resist temperature changes. For example, when you heat an iron kettle, it gets hot pretty quickly. Why? Because it has a *low* specific heat. It doesn't take much heat to increase the temperature of the kettle. Water, on the other hand, has a high heat capacity. You have to add a lot of heat to get an increase in temperature. The boiling point of water is pretty darn high. Liquid water is essential for cellular structures, and it is important that it doesn't boil away if someone goes out on a warm day. Water's ability to resist temperature changes is one of the things that helps keep the temperature in our oceans fairly stable. It's also why organisms that are mainly made up of water, like us, are able to keep a constant body temperature.

Expansion on Freezing

One of the most important properties of water is yet another result of hydrogen bonding. When four water molecules are bound in a solid lattice of ice, the hydrogen bonds actually cause solid water to expand on freezing. In most materials, when the molecules lose kinetic energy and cool from liquid to solid, the molecules get more dense, moving closer together. However, in liquid water, the molecules are slightly more dense than in solid water. Because water expands on freezing, becoming slightly less dense than liquid water, ice can crack lead pipes in the winter, or a soda can will pop if it is left in your cold car. The important consequence of this property of water is that ice floats on the top of lakes or streams, allowing animal life to live underneath the ice. If ice was denser than water, it would sink to the bottom, freezing the body of water solid. All aquatic life would be unable to survive.

So let's review the unique properties of water:

- Water is polar and can dissolve other polar substances.
- Water has cohesive and adhesive properties.
- Water has a high heat capacity.
- Water expands when it freezes (or, ice floats).

ACIDS AND BASES

We just said that water is important because most reactions occur in watery solutions. Well, there's one more thing to remember: reactions are also influenced by whether the solution in which they occur is **acidic**, **basic**, or **neutral**.

What makes a solution acidic or basic? A solution is acidic if it contains a lot of *hydrogen ions* (H^+). That is, if you dissolve an acid in water, it will release a lot of hydrogen ions. When you think about acids, you usually think of substances with a sour taste, like lemons. For example, if you squeeze a little lemon juice into a glass of water, the solution will become acidic. That's because lemons contain citric acid, which releases a lot of H^+ into the solution.

Important Formulas
Don't forget to check out what formulas will be given on the equations and formulas sheet.

Bases, on the other hand, do not release hydrogen ions when added to water. They release a lot of *hydroxide ions* (OH⁻). These solutions are said to be **alkaline** (the fancy name for a basic solution). Bases usually have a slippery consistency. Common soap, for example, is composed largely of bases.

The acidity or alkalinity of a solution can be measured with a **pH scale**. The pH scale is numbered from 1 to 14. The midpoint, 7, is considered neutral pH. The concentration of hydrogen ions in a solution will indicate whether it is acidic, basic, or neutral. If a solution contains a lot of hydrogen ions, then it will be acidic and have a low pH. Here's the trend:

> An increase in H⁺ ions causes a decrease in the pH.
>
> $$pH = -\log [H^+]$$

One more thing to remember: the pH scale is not a linear scale—it's logarithmic. That is, a change of *one* pH number actually represents a *tenfold* change in hydrogen ion concentration. For example, a pH of 3 is actually ten times more acidic than a pH of 4. This is also true in the reverse direction: a pH of 4 represents a tenfold decrease in acidity compared to a pH of 3.

Therefore, as the concentration of H⁺ ions increases by a factor of 10, the pH becomes one number smaller. For example, stomach acid has a pH of 2, and if we use the equation, we discover that the concentration of H⁺ ions in stomach acid is 10^{-2} *M*. This is pretty high when you consider the other extreme is lye, which has a pH of 14, a concentration of H⁺ ions around 10^{-14} *M*! Use your calculator to double-check that these numbers are correct.

You'll notice from the scale below that stronger acids have lower pHs. If a solution has a low concentration of hydrogen ions, it will have a high pH.

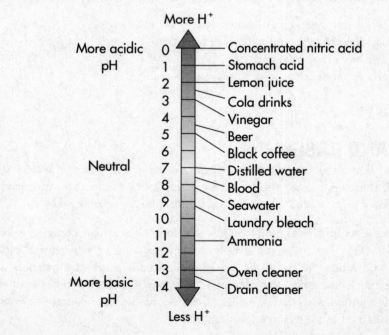

The equation for pH is listed on the AP Biology Equations and Formulas sheet. However, you will not be expected to perform calculations using this equation. Instead, you should understand how the equation works and when pH calculations are useful.

ORGANIC MOLECULES

Now that we've discussed chemical compounds in general, let's talk about a special group of compounds. Most of the chemical compounds in living organisms contain a skeleton of carbon atoms surrounded by hydrogen atoms and often other elements. These molecules are known as **organic compounds**. By contrast, molecules that do not contain carbon atoms are called **inorganic compounds**. For example, salt (NaCl) is an inorganic compound.

Carbon is important for life because it is a versatile atom, meaning that it has the ability to bind not only with other carbons but also with a number of other elements including nitrogen, oxygen, and hydrogen. The resulting molecules are key in carrying out the activities necessary for life.

To recap:

- Organic compounds contain carbon atoms AND hydrogen atoms (and sometimes other things too).
- Inorganic compounds don't contain both carbon atoms and hydrogen atoms.

Now let's focus on the following four classes of organic compounds central to life on Earth:

- carbohydrates
- proteins
- lipids
- nucleic acids

Most macromolecules are chains of building blocks, called **polymers**. The individual building blocks of a polymer are called **monomers**.

Carbohydrates

Organic compounds that contain carbon, hydrogen, and oxygen are called **carbohydrates**. They usually contain these three elements in a ratio of appoximately 1:2:1, respectively. We can represent the proportion of elements within carbohydrate molecules by the formula $C_nH_{2n}O_n$, which can be simplified as $(CH_2O)_n$.

Most carbohydrates are categorized as either **monosaccharides**, **disaccharides**, or **polysaccharides**. The term *saccharides* is a fancy word for "sugar." The prefixes *mono-*, *di-*, and *poly-* refer to the number of sugars in the molecule. *Mono-* means "one," *di-* means "two," and *poly-* means "many." A monosaccharide is therefore a carbohydrate made up of a single saccharide.

Monosaccharides: The Simplest Sugars

Monosaccharides, the simplest sugars, serve as an energy source for cells. The two most common sugars are (1) **glucose** and (2) **fructose**.

Both glucose and fructose are six-carbon sugars with the chemical formula $C_6H_{12}O_6$. Glucose, the most abundant monosaccharide, is the most popular sugar around. Glucose is an important part of the food we eat and it is the favorite food made by plants during photosynthesis. This glucose can be broken down to give us (and plants) energy. Fructose, the other monosaccharide you need to know for the test, is a common sugar in fruits.

Glucose and fructose can be depicted as either "straight" or "rings." Both of them are pretty easy to spot; just look for the carbon molecules with a lot of OHs and Hs attached to them.

Here are the two different forms.

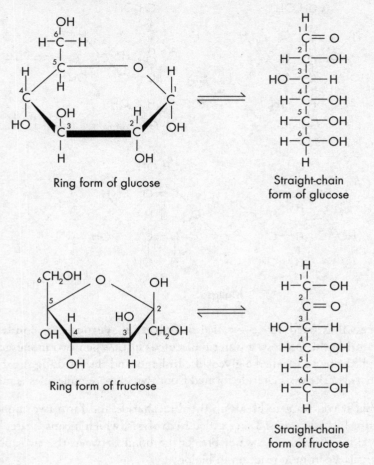

Ring form of glucose

Straight-chain form of glucose

Ring form of fructose

Straight-chain form of fructose

Disaccharides

What happens when two monosaccharides are brought together? The hydrogen (–H) from one sugar molecule combines with the hydroxyl group (–OH) of another sugar molecule. What do H and OH add up to? Water (H_2O)!

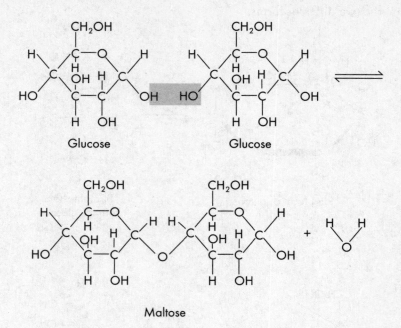

Glucose   Glucose

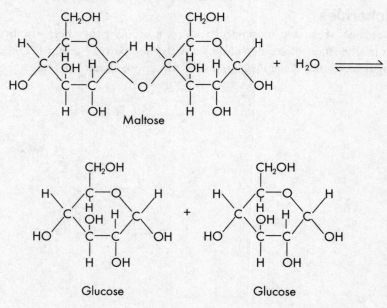

Maltose

The process illustrated above is called **dehydration synthesis**, or **condensation**, because during this process, a water molecule is lost. When two monosaccharides are joined, the bond is called a **glycosidic linkage,** and the resulting sugar is called a disaccharide. The disaccharide formed from two glucose molecules is maltose.

Now, what if you want to break up the disaccharide and form two monosaccharides again? Just add water. That's called **hydrolysis**, which means "water" (*hydro-*) and "breaking" (*-lysis*). The water breaks the bond between the two monomers. Hydrolysis is a common reaction in biology.

Maltose

Glucose   Glucose

Polysaccharides

Polysaccharides are made up of many repeated units of monosaccharides. The most common polysaccharides you'll need to know for the test are **starch**, **cellulose**, and **glycogen**. Glycogen and starch are sugar storage molecules. Glycogen stores sugar in animals and starch stores sugar in plants. Cellulose, on the other hand, is made up of β-glucose and is a major part of the cell walls in plants. Its function is to lend structural support. Chitin, a polymer of β-glucose molecules, serves as a structural molecule in the walls of fungus and in the exoskeletons of arthropods.

Proteins

Amino acids are organic molecules that serve as the building blocks of proteins. They contain carbon, hydrogen, oxygen, and nitrogen atoms. There are 20 different amino acids commonly found in proteins. Fortunately, you don't have to memorize the 20 amino acids. But you do have to remember that every amino acid has four important parts around a central carbon: an **amino group** (–NH$_2$), a **carboxyl group** (–COOH), a hydrogen, and an **R group**.

Here's a typical amino acid.

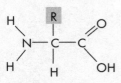

Amino acids differ only in the R group, which is also called the **side chain**. The R group associated with an amino acid could be as simple as a hydrogen atom (as in the amino acid *glycine*) or as complex as a charged carbon skeleton (as in the amino acid *arginine*).

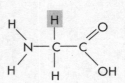

Glycine

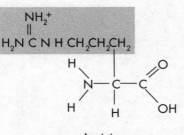

Arginine

When it comes to spotting an amino acid, simply keep an eye out for the amino group (NH$_2$); then look for the carboxyl molecule (COOH). The side chains for each amino acid vary greatly. You don't have to memorize the structures of the side chains for individual amino acids, but you should know that they can vary in composition, polarity, charge, and shape depending on the side chain that they have.

How do the side chains of the amino acids differ?

- Composition of elements (C, H, O, N, and S)
- Polarity (polar, nonpolar)
- Charge (neutral, positive, negative)
- Shape (long chain, short chain, ring-shape)

Naturally Occurring Amino Acids

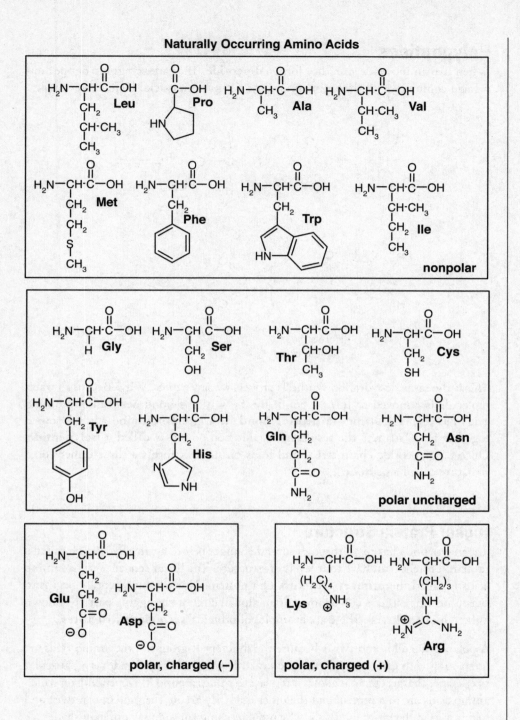

Polypeptides

When two amino acids join, they form a **dipeptide**. The carboxyl group of one amino acid combines with the amino group of another amino acid. Here's an example:

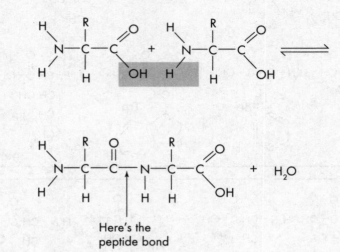

Here's the
peptide bond

This is the same dehydration synthesis process we saw earlier. Why? Because a water molecule is removed to form a bond. By the way, the bond between two amino acids has a special name—a **peptide bond**. If a group of amino acids is joined together in a "string," the resulting organic compound is called a **polypeptide**. Once a polypeptide chain twists and folds on itself, it forms a three-dimensional structure called a **protein**.

Higher Protein Structure

The polypeptide has to go through several changes before it can officially be called a protein. Proteins can have four levels of structure. The linear sequence of the amino acids is called the **primary structure** of a protein. Now the polypeptide begins to twist, forming either a coil (known as an alpha helix) or zigzagging pattern (known as beta-pleated sheets). These are examples of proteins' **secondary structures**.

A polypeptide folds and twists because the different R-groups of the amino acids are interacting with each other. Remember, each R-group is unique and has a particular size, shape, charge, etc. that allows it to react to things around it. Depending on which amino acids are in a protein and the order that they are in, the protein can twist and fold in very different ways. This is why proteins can form so many different shapes.

The secondary structure is formed by amino acids that interact with other amino acids closeby in the primary structure. However, after the secondary structure reshapes the polypeptide, amino acids that were far away in the primary structure arrangement can now also interact with each other. This is called the **tertiary structure**. Sometimes, two cysteine amino acids can react with each other to form a covalent disulfide bond that stabilizes the tertiary structure.

Lastly, several different polypeptide chains sometimes interact with each other to form a **quaternary structure**. The different polypeptide chains that come together are often called subunits of the final whole protein.

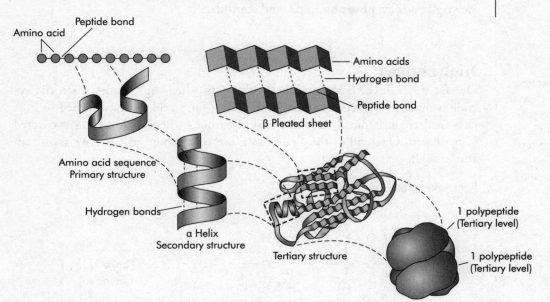

Only proteins that have folded correctly into their specific three-dimensional structure can perform their function properly. Mistakes in the amino acid chain can create oddly shaped proteins that are nonfunctional. One more thing: in some cases, the folding of proteins involves other proteins known as **chaperone proteins** (or **chaperonins**). They help the protein fold properly and make the process more efficient.

Lipids

Like carbohydrates, **lipids** consist of carbon, hydrogen, and oxygen atoms, but not in the 1:2:1 ratio typical of carbohydrates. The most common examples of lipids are **triglycerides, phospholipids**, and **steroids.**

Triglycerides

Make the Connection
Amino acids have carboxyl groups and fatty acids have carboxyl groups. Carboxyl groups are acidic; hence they both have acid in their name.

Our fat stores in the body are filled with lipids called triglycerides. Each triglyceride is made of a glycerol molecule with three fatty acid chains attached to it. A fatty acid chain is mostly a long chain of carbons where each carbon is covered in hydrogen. One end of the chain has a carboxyl group (just like we saw in an amino acid).

Let's take a look.

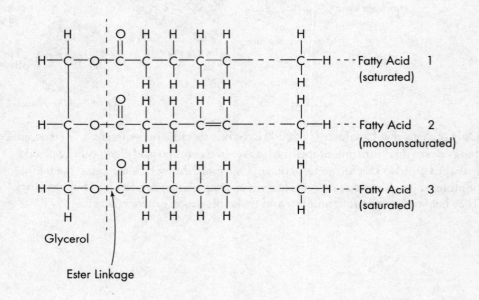

To make a triglyceride, each of the carboxyl groups (–COOH) of the three fatty acids must react with one of the three hydroxyl groups (–OH) of the glycerol molecule. This happens by the removal of a water molecule. So, the creation of a fat requires the removal of three molecules of water. Once again, what have we got? You probably already guessed it—dehydration synthesis! The linkage now formed between the **glycerol** molecule and the fatty acids is called an **ester linkage**. A fatty acid can be **saturated** with hydrogens along its long carbon chain or it can have a few gaps where double bonds exist instead of a hydrogen. It there is a double bond in the chain, then it is an **unsaturated** fatty acid. A **polyunsaturated** fatty acid has many double bonds within the fatty acid.

Lipids are important because they function as structural components of cell membranes, sources of insulation, and a means of energy storage.

Phospholipids

Another special class of lipids is known as phospholipids. Phospholipids contain two fatty acid "tails" and one negatively charged phosphate "head." Take a look at a typical phospholipid.

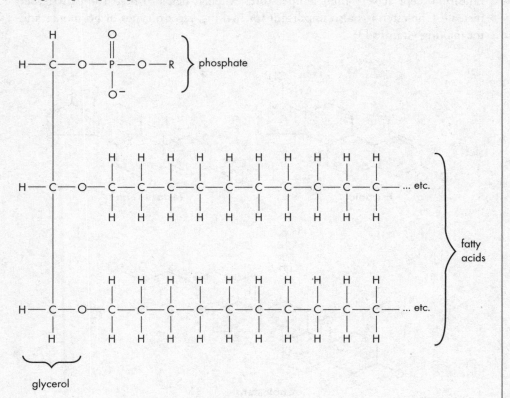

Phospholipids are extremely important, mainly because of some unique properties they possess, particularly with regard to water.

The two fatty acid tails are **hydrophobic** ("water-hating"). In other words, just like oil and vinegar, fatty acids and water don't mix. The reason for this is that fatty acid tails are nonpolar, and nonpolar substances don't mix well with polar ones, such as water.

On the other hand, the phosphate "head" of the lipid is **hydrophilic** ("water-loving"), meaning that it does mix well with water. Why? It carries a negative charge, and this charge draws it to the positively charged end of a water molecule. Since a phospholipid has both a hydrophilic region and a hydrophobic region, it is an **amphipathic molecule**. One side of a phospholipid loves to hang out with water, and the other side hates to.

Thus, the two fatty acid chains orient themselves away from water, while the phosphate portion orients itself toward the water. Keep these properties in mind. We'll see later how this orientation of phospholipids in water relates to the structure and function of cell membranes.

Cholesterol

Cholesterol is another important type of lipid. Cholesterol is a four-ringed molecule that is found here and there in membranes. It generally increases membrane fluidity, except at very high temperatures when it helps to hold things together instead. Cholesterol is also important for making certain types of hormones and for making vitamin D.

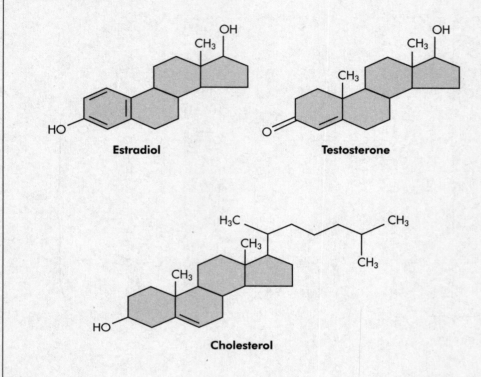

Estradiol

Testosterone

Cholesterol

Nucleic Acids

The fourth class of organic compounds is the **nucleic acids.** Like proteins, nucleic acids contain carbon, hydrogen, oxygen, and nitrogen, but nucleic acids also contain phosphorus. Nucleic acids are molecules that are made up of simple units called **nucleotides.** For the AP Biology Exam, you'll need to know about two kinds of nucleic acids: **deoxyribonucleic acid (DNA)** and **ribonucleic acid (RNA).**

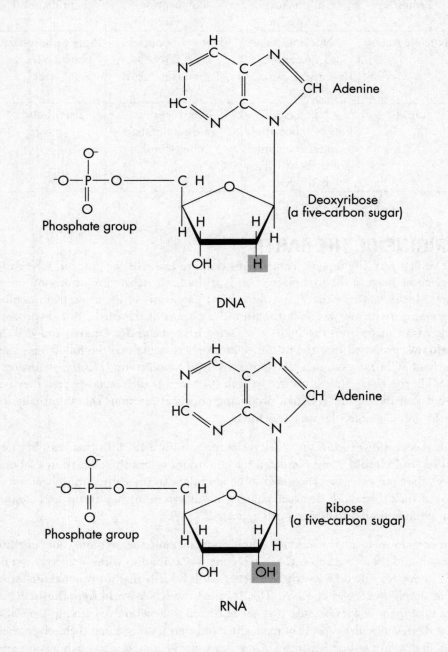

Need-to-Know Nucleic Acids
For the AP Biology Exam, you'll be expected to know about two types of nucleic acids, DNA and RNA, which are shown here.

DNA is important because it contains the hereditary "blueprints" of all life. RNA is important because it's essential for protein synthesis. We'll discuss DNA and RNA in greater detail when we discuss heredity. The following table summarizes the important macromolecules discussed here.

Important Biological Macromolecules

Macromolecule	Monomer	Polymer	Linkage Bond
Carbohydrates	Monosaccharide (e.g. Glucose)	Polysaccharide (e.g. Starch, glycogen, cellulose)	Glycosidic linkage
Proteins	Amino Acid (e.g. Glycine)	Polypeptide (e.g. Actin)	Peptide bond
Nucleic Acids	Nucleotides (e.g. Adenine, thymine, guanine, cytosine)	Deoxyribonucleic acid (DNA) Ribonucleic acid (RNA)	Sugar-phosphate phosphodiester bonds
Lipids	Not a true polymer, but often contains chains of carbons with hydrogens	Triglycerides, phospholipids, cholesterol	Ester bonds

ORIGINS OF THE EARTH

We've just seen the organic compounds that are essential for life. But where did they come from in the first place? This is still a hotly debated topic among scientists. Most scientists believe that the earliest precursors of life arose from nonliving matter (basically gases) in the primitive oceans of the earth. But this theory didn't take shape until the 1920s. Two scientists, **Alexander Oparin** and **J. B. S. Haldane**, proposed that the primitive atmosphere contained the following gases: methane (CH_4), ammonia (NH_3), hydrogen (H_2), and water (H_2O). Interestingly enough, there was almost no free oxygen (O_2) in this early atmosphere. They believed that these gases collided, producing chemical reactions that eventually led to the organic molecules we know today.

This theory didn't receive any substantial support until 1953. In that year, **Stanley Miller** and **Harold Urey** simulated the conditions of primitive Earth in a laboratory. They put the gases theorized to be abundant in the early atmosphere into a flask, struck them with electrical charges in order to mimic lightning, and organic compounds similar to amino acids appeared!

But how do we make the leap from simple organic molecules to more complex compounds and life as we know it? Since no one was around to witness the process, no one knows for sure how (or when) it occurred. It is likely that the original life-forms were simply molecules of RNA. This is called the **RNA-world hypothesis**. RNA can take many shapes because it is not restricted to be a double helix. It is possible that RNA molecules capable of replicating and thus passing along themselves (their genome) were the first life-forms. Complex organic compounds (such as proteins) must have formed via dehydration synthesis. Simple cells then used organic molecules as their source of food. Over time, simple cells evolved into complex cells.

Now let's throw in a few new terms. Living organisms that rely on organic molecules for food are called **heterotrophs** (or consumers). For example, we're heterotrophs, but the earliest heterotrophs were simple unicellular life-forms. Eventually, some life-forms found a way to make their own food—most commonly through photosynthesis. These organisms are called **autotrophs** (or producers). Early autotrophs (most likely cyanobacteria) are responsible for Earth's oxygenated atmosphere.

KEY TERMS

elements
oxygen
carbon
hydrogen
nitrogen
trace elements
atom
protons
neutrons
electrons
nucleus
isotopes
radiometric dating
compound
chemical reaction
chemical bond
ionic bond
covalent bond
hydrogen bond
ions
nonpolar covalent
polar covalent
polar
cohesion
adhesion
heat capacity
expansion on freezing
capillary action
acidic
basic
neutral
alkaline
pH scale
organic compounds
inorganic compounds
polymer
monomer
carbohydrates
monosaccharides
disaccharides
polysaccharides
glucose
fructose

dehydration synthesis
  (or condensation)
glycosidic linkage
hydrolysis
starch
cellulose
glycogen
amino acids
amino group
carboxyl group
R group
side chain
dipeptide
peptide bond
polypeptide
protein
primary structure
secondary structure
tertiary structure
quaternary structure
chaperone proteins (chaperonins)
lipids
triglycerides
phospholipids
steroids
glycerol
ester linkage
saturated
unsaturated
polyunsaturated
hydrophobic
hydrophilic
amphipathic
cholesterol
nucleic acids
nucleotides
deoxyribonucleic acid (DNA)
ribonucleic acid (RNA)
Alexander Oparin and J. B. S. Haldane
Stanley Miller and Harold Urey
RNA-world hypothesis
heterotrophs (or consumers)
autotrophs (or producers)

Summary

○ Compounds are composed of two or more types of elements in a fixed ratio.

○ Molecules are held together by ionic bonds and covalent bonds.

○ Hydrogen bonding is essential for the following emergent properties of water:
   • cohesion
   • adhesion
   • surface tension
   • high heat capacity
   • expansion upon freezing
   • polar structure
   • ability to dissolve other polar (hydrophilic) substances but inability to dissolve nonpolar (hydrophobic) substances

○ pH is important for biological reactions. Solutions may be acidic, basic or alkaline, or neutral.

○ Organic molecules contain carbon atoms. Many molecules in biology are composed of several monomers bound together into polymers. The creation of these polymers occurs through dehydration synthesis reactions. The breakdown of these polymers occurs through hydrolysis reactions.

○ Important organic macromolecules in biology include
   • Carbohydrates: monosaccharides link by glyosidic bonds to form di- and polysaccharides. Examples are glucose, fructose, glycogen, cellulose, and starch.
   • Proteins: 20 different amino acids, each with a different R-group, link together via peptide bonds to form polypeptides. The chain further folds up into special shapes to make the final protein structure. The structure can be divided into primary (amino acid sequence), secondary (alpha helices and beta pleated sheets), tertiary (disulfide bonds), and quaternary (multiple polypeptides).
   • Lipids: chains called fatty acids (either saturated or unsaturated) commonly form as phospholipids and triglycerides. Cholesterol is another type of lipid. Lipids are hydrophobic.
   • Nucleic acids: chains of either DNA or RNA nucleotides. They carry the genetic recipes in the body.

○ Heterotrophs are consumers and autotrophs are producers.

Chapter 4 Drill

Answers and explanations can be found in Chapter 16.

1. Water is a critical component of life due to its unique structural and chemical properties. Which of the following does NOT describe a way that the exceptional characteristics of water are used in nature to sustain life?

   (A) The high heat capacity of water prevents lakes and streams from rapidly changing temperature and freezing completely solid in the winter.
   (B) The high surface tension and cohesiveness of water facilitates capillary action in plants.
   (C) The low polarity of water prevents dissolution of cells and compounds.
   (D) The high intermolecular forces of water, such as hydrogen bonding, result in a boiling point which exceeds the tolerance of most life on the planet.

2. The intracellular pH of human cells is approximately 7.4. Yet, the pH within the lumen (inside) of the human stomach averages 1.5. Which of the following accurately describes the difference between the acidity of the cellular and gastric pH?

   (A) Gastric juices contain approximately 5 times more H^+ ions than the intracellular cytoplasm of cells and are more acidic.
   (B) Gastric juices contain approximately 100,000-fold more H^+ ions than the intracellular cytoplasm of cells and are more acidic.
   (C) The intracellular cytoplasm of cells contain approximately 5 times more H^+ ions than gastric juices and is more acidic.
   (D) The intracellular cytoplasm of cells contains approximately 100,000-fold more H^+ ions than gastric juices and is more acidic.

3. Amino acids are the basic molecular units which compose proteins. All life on the planet forms proteins by forming chains of amino acids. Which labeled component of the amino acid structure of phenylalanine shown below will vary from amino acid to amino acid?

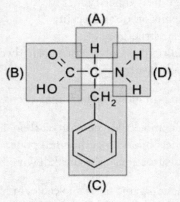

Questions 4-6 refer to the following paragraph and diagram.

In 1953, Stanley Miller and Harold Urey performed an experiment at the University of Chicago to test the hypothesis that the conditions of the early Earth would have favored the formation of larger, more complex organic molecules from basic precursors. The experiment, as shown below, consisted of sealing basic organic chemicals (representing the atmosphere of the primitive Earth) in a flask, which was exposed to electric sparks (to simulate lightning) and water vapor.

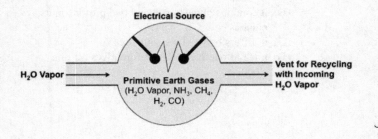

After one day of exposure, the mixture in the flask had turned pink in color, and later analysis showed that at least 10% of the carbon had been transformed into simple and complex organic compounds including at least 11 different amino acids and some basic sugars. No nucleic acids were detected in the mixture.

4. Which of the following contradicts the hypothesis of the experiment that life may have arisen from the formation of complex molecules in the conditions of the primitive Earth?

    (A) Complex carbon-based compounds were generated after only one day of exposure to simulated primitive Earth conditions.
    (B) Nucleic acid compounds such as DNA and RNA were not detected in the mixture during the experiment.
    (C) Over half of known amino acids involved in life were detected in the mixture during the experiment.
    (D) Basic sugar molecules were generated and detected in the mixture during the experiment.

5. Some amino acids, such as cysteine (shown below) and methionine, could not be formed in this experiment. Which of the following best explains why these molecules could not be detected?

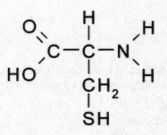

    (A) The chemical reactions necessary to create amino acids such as cysteine and methionine require more energy than the simulated lightning provided in the experiment.
    (B) The chemical reactions necessary to create amino acids such as cysteine and methionine require enzymes for catalysis to occur, which were not included in the experiment.
    (C) Sulfur-based compounds were not included in the experiment.
    (D) Nitrogen-based compounds were not included in the experiment.

6. A scientist believes that the Miller-Urey experiment failed to yield the remaining amino acids and the nucleic acids because of the absence of critical chemical substrates that would have existed on the primordial Earth due to volcanism. Which of the following basic compounds, which are associated with volcanism, would not need to be added in a follow-up Miller-Urey experiment?

    (A) H_2S gas
    (B) SiO_2 (silica)
    (C) SO_2
    (D) H_3PO_4 (phosphoric acid)

7. Catabolism refers to breaking down complex macromolecules into their basic components. Many biological processes use hydrolysis for catabolism. Hydrolysis of proteins could directly result in

(A) free water
(B) adenine
(C) cholesterol
(D) dipeptides

8. Which of the following contain both hydrophilic and hydrophobic properties and are often found in cell plasma membranes?

(A) Nucleotides
(B) Phospholipids
(C) Water
(D) Amino acids

9. Maltotriose is a trisaccharide composed of three glucose molecules linked through α-1,4 glycosidic linkages formed via dehydration synthesis. What would the formula be for maltotriose?

(A) $C_{18}H_{36}O_{18}$
(B) $C_{18}H_{10}O_{15}$
(C) $C_{18}H_{32}O_{16}$
(D) $C_3H_6O_3$

10. A radioactive isotope of hydrogen, ^3H, is called tritium. Tritium differs from the more common form of hydrogen because

(A) it contains two neutrons and one proton in its nucleus
(B) it contains one neutron and two protons in its nucleus
(C) it differs by its atomic number
(D) it is radioactive and therefore gives off one electron

REFLECT

Respond to the following questions:

- Which topics from this chapter do you feel you have mastered?

- Which content topics from this chapter do you feel you need to study more before you can answer multiple-choice questions correctly?

- Which content topics from this chapter do you feel you need to study more before you can effectively compose a free response?

- Was there any content that you need to ask your teacher or another person about?

Chapter 5
Cells

LIVING THINGS

All living things are composed of **cells**. According to the cell theory, the cell is life's basic unit of structure and function. This simply means that the cell is the smallest unit of living material that can carry out all the activities necessary for life.

It may seem strange that we are still made of tiny little cells. Why not just be a living thing that is one giant cell?

One reason for this is because compartments allow more specialization. As we will see, compartmentalization is an important part of the organization of unicellular and multicellular organisms.

The second reason we can't just be giant cells is the surface area-to-volume ratio. There is always lots of exchange going on between the inside of things and the outside. This ratio must be kept large so that there is lots of space to do these exchanges.

Therefore, even the largest things are still made of cells.

Surface area-to-volume ratio

This is a concept that comes up often. Use the volume formulas on the Equations and Formulas sheet to practice calculating the surface area-to-volume ratios of different-sized cubes and spheres.

MICROSCOPES AND THE STUDY OF CELLS

Cells are studied using different types of microscopes. **Light microscopes** (for example, the compound microscopes commonly found in labs) are used to study stained or living cells. They can magnify the size of an organism up to 1,000 times. **Electron microscopes** are used to study detailed structures of a cell that cannot be easily seen or observed by light microscopy. They are capable of resolving structures as small as a few nanometers in length, such as individual virus particles or the pores on the surface of the nucleus.

TYPES OF CELLS AND ORGANELLES

For centuries, scientists have known about cells. However, it wasn't until the development of the electron microscope that scientists were able to figure out what cells do. We now know that there are two distinct types of cells: **prokaryotic cells** and **eukaryotic cells**.

A prokaryotic cell, which is a lot smaller than a eukaryotic cell, is relatively simple. Bacteria, for example, are prokaryotic cells. The inside of the cell is filled with a substance called **cytoplasm**. The genetic material in a prokaryote is one continuous, circular DNA molecule that is found free in the cell in an area called the **nucleoid** (this is not the same as a nucleus!). Most prokaryotes have a **cell wall** composed of peptidoglycans that surrounds a lipid layer called the **plasma membrane**. Prokaryotes also have ribosomes (though smaller than those found in eukaryotic cells). Some bacteria may also have one or more **flagella**, which are long projections used for motility (movement) and they might have a thick **capsule** outside their cell wall to give them extra protection.

Eukaryotic cells are more complex than prokaryotic cells. Fungi, protists, plants, and animals are eukaryotes. Eukaryotic cells are organized into many smaller structures called **organelles**. Some of these organelles are the same structures seen in prokaryotic cells, but many are uniquely eukaryotic. A good way to remember the difference is that prokaryotes do not have any membrane-bound organelles. Their only membrane is the plasma membrane.

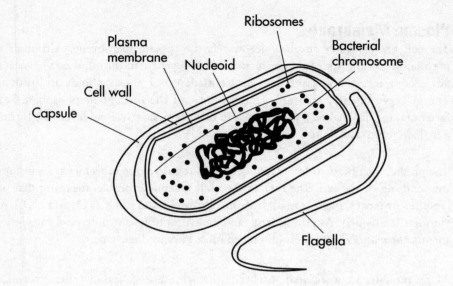

Organelles

A eukaryotic cell is like a microscopic factory. It's filled with organelles, each of which has its own special tasks. Let's take a tour of a eukaryotic cell and focus on the structure and function of each organelle. Here's a picture of a typical animal cell and its principal organelles.

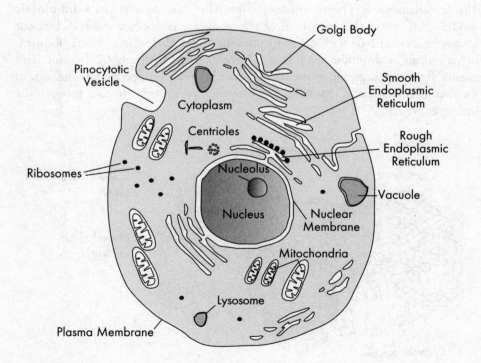

Plasma Membrane

The cell has an outer envelope known as the plasma membrane. Although the plasma membrane appears to be a simple thin layer surrounding the cell, it's actually a complex, double-layered structure made up of phospholipids and proteins. The *hydrophobic* fatty acid tails face inward and the *hydrophilic* phosphate heads face outward. It is called a **phospholipid bilayer** since two lipid layers are forming a hydrophobic sandwich.

The plasma membrane is important because it regulates the movement of substances into and out of the cell. The membrane itself is semipermeable, meaning that only certain substances, namely small hydrophobic molecules (such as O_2 and CO_2), pass through it unaided. Anything large and/or hydrophilic can only pass through the membrane via special tunnels (discussed more later in this chapter).

Many proteins are associated with the cell membrane. Some of these proteins are loosely associated with the lipid bilayer (**peripheral proteins**). They are located on the inner or outer surface of the membrane. Others are firmly bound to the plasma membrane (**integral proteins**). These proteins are amphipathic, which means that their hydrophilic regions extend out of the cell or into the cytoplasm, while their hydrophobic regions interact with the tails of the membrane phospholipids. Some integral proteins extend all the way through the membrane (**transmembrane proteins**).

This arrangement of phospholipids and proteins is known as the **fluid-mosaic model**. This means that each layer of phospholipids is flexible, and it is a mosaic because it is peppered with different proteins and carbohydrate chains. Remember, anything hydrophilic should not go through the hydrophobic interior. This means that the phospholipids on one side should never flip-flop to the other side of the membrane (because that would require their polar heads to pass through the hydrophobic area).

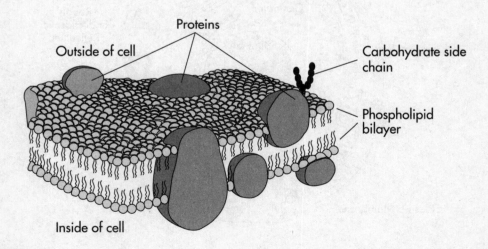

Why should the plasma membrane need so many different proteins? It's because of the number of activities that take place in or on the membrane. Generally, plasma membrane proteins fall into several broad functional groups. Some membrane proteins form junctions between adjacent cells (**adhesion proteins**). Others serve as docking sites for arrivals at the cell, such as hormones (**receptor proteins**). Some proteins form pumps that use ATP to actively transport solutes across the membrane (**transport proteins**). Others form channels that selectively allow the passage of certain ions or molecules (**channel proteins**). Finally, some proteins, such as glycoproteins, are exposed on the extracellular surface and play a role in cell recognition and adhesion (**cell surface markers**).

Attached to the surface of some proteins are **carbohydrate side chains**. They are found only on the outer surface of the plasma membrane. As mentioned above, cholesterol molecules are also found in the phospholipid bilayer because they help stabilize membrane fluidity in animal cells.

The Nucleus

The **nucleus** is usually the largest organelle in the cell. The nucleus not only directs what goes on in the cell, it is also responsible for the cell's ability to reproduce. It's the home of the hereditary information—DNA—which is organized into large structures called **chromosomes**. The most visible structure within the nucleus is the **nucleolus**, which is where rRNA is made and ribosomes are assembled.

Ribosomes

The **ribosomes** are the sites of protein synthesis. Their job is to manufacture all the proteins required by the cell or secreted by the cell. Ribosomes are round structures composed of two subunits, the large subunit and the small subunit. The structure is composed of RNA and proteins. Ribosomes can be either free floating in the cell or attached to another structure called the **endoplasmic reticulum (ER)**.

Endoplasmic Reticulum (ER)

The endoplasmic reticulum (ER) is a continuous channel that extends into many regions of the cytoplasm. The region of the ER that is attached to the nucleus and "studded" with ribosomes is called the **rough ER** (RER). Proteins generated in the rough ER are trafficked to or across the plasma membrane, or they are used to build Golgi bodies, lysosomes, or the ER. The region of the ER that lacks ribosomes is called the **smooth ER** (SER). The smooth ER makes lipids, hormones, and steroids and breaks down toxic chemicals.

Golgi Bodies

The **Golgi bodies**, which look like stacks of flattened sacs, also participate in the processing of proteins. Once the ribosomes on the rough ER have completed synthesizing proteins, the Golgi bodies modify, process, and sort the products. They're the packaging and distribution centers for materials destined to be sent out of the cell. They package the final products in little sacs called **vesicles**, which carry the products to the plasma membrane. Golgi bodies are also involved in the production of lysosomes.

Mitochondria

Another important organelle is the mitochondrion. **Mitochondria** are often referred to as the "powerhouses" of the cell. They're power stations responsible for converting the energy from organic molecules into useful energy for the cell. The energy molecule in the cell is **adenosine triphosphate (ATP)**.

The mitochondrion is usually an easy organelle to recognize because it has a unique oblong shape and a characteristic double membrane consisting of an inner portion and an outer portion. The inner mitochondrial membrane forms folds known as **cristae** and separates the innermost area called the matrix from the intermembrane space. The outer membrane separates the intermembrane space from the cytoplasm. As we'll see later, most of the production of ATP is done on the cristae.

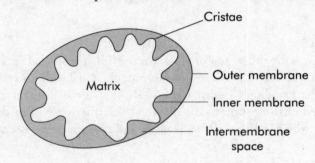

Lysosomes

Throughout the cell are small, membrane-enclosed structures called **lysosomes**. These tiny sacs carry digestive enzymes, which they use to break down old, worn-out organelles, debris, or large ingested particles. Lysosomes make up the cell's clean-up crew, helping to keep the cytoplasm clear of unwanted flotsam. Lysosomes contain hydrolytic enzymes that function only at acidic pH, which is enclosed inside the lumen of the lysosome.

Centrioles

The **centrioles** are small, paired, cylindrical structures that are often found within **microtubule organizing centers (MTOCs)**. Centrioles are most active during cellular division. When a cell is ready to divide, the centrioles produce microtubules, which pull the replicated chromosomes apart and move them to opposite ends of the cell. Although centrioles are common in animal cells, they are not found in most types of plant cells.

Vacuoles

In Latin, the term *vacuole* means "empty cavity." But **vacuoles** are far from empty. They are fluid-filled sacs that store water, food, wastes, salts, or pigments.

Peroxisomes

Peroxisomes are organelles that detoxify various substances, producing hydrogen peroxide (H_2O_2) as a byproduct. They also contain enzymes that break down hydrogen peroxide into oxygen and water. In animals, they are common in the liver and kidney cells.

Cytoskeleton

Have you ever wondered what actually holds the cell together and enables it to keep its shape? The shape of a cell is determined by a network of fibers called the **cytoskeleton**. The most important fibers you'll need to know are **microtubules** and **microfilaments**.

Microtubules, which are made up of the protein **tubulin**, participate in cellular division and movement. These small fibers are an integral part of three structures: **centrioles, cilia,** and flagella. We've already mentioned that centrioles help chromosomes separate during cell division.

Microfilaments, like microtubules, are important for movement. These thin, rod-like structures are composed of the protein actin. Actin monomers are joined together and broken apart as needed to allow microfilaments to grow and shrink. Microfilaments assist during cytokinesis, muscle contraction, and formation of pseudopodia extensions during cell movement.

Cilia and Flagella

Cilia and flagella are threadlike structures best known for their locomotive properties in single-celled organisms. The beating motion of cilia and flagella structures propels these organisms through their watery environments.

The two classic examples of organisms with these structures are the *Euglena*, which gets about using its whiplike flagellum, and the *Paramecium*, which is covered in cilia. The rhythmic beating of the *Paramecium*'s cilia enables it to motor about in waterways, ponds, and microscope slides in your biology lab. You may have seen these in lab, but here's what they look like.

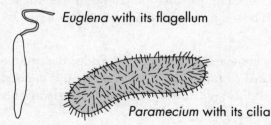

Euglena with its flagellum

Paramecium with its cilia

Though we usually associate such structures with microscopic organisms, they aren't the only ones with cilia and flagella. As you probably know, these structures are also found in certain human cells. For example, the cells lining your respiratory tract possess cilia that sweep constantly back and forth (beating up to 20 times per second), helping to keep dust and unwanted debris from descending into your lungs. And every sperm cell has a type of flagellum, which enables it to swim through the female reproductive organs to fertilize the waiting ovum.

Plant Cells Versus Animal Cells

Plant cells contain most of the same organelles and structures seen in animal cells, with several key exceptions. Plant cells, unlike animal cells, have a protective outer covering called the **cell wall** (made of cellulose). A cell wall is a rigid layer just outside of the plasma membrane that provides support for the cell. It is found in plants, protists, fungi, and bacteria. (In fungi, the cell wall is usually made of **chitin**, a modified polysaccharide. Chitin is also a principle component of an arthropod's exoskeleton.) In addition, plant cells possess **chloroplasts**. Chloroplasts contain chlorophyll, the light-capturing pigment that gives plants their characteristic green color. Because chloroplasts are involved in photosynthesis, we will discuss them in more detail in Chapter 6. Another difference between plant and animal cells is that most of the cytoplasm within a plant cell is usually taken up by a large vacuole that crowds the other organelles. In mature plants, this vacuole contains the **cell sap**. A full vacuole in a plant is a sign that it is not dehydrated. Dehydrated plants cannot fill their vacuoles and they wilt. Plant cells also differ from animal cells in that plant cells do not contain centrioles.

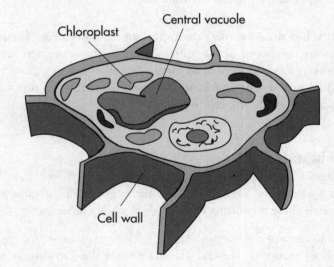

To help you remember the differences among prokaryotes, plant cells, and animal cells, we've put together a simple table (on the next page). Make sure you learn it! The testing board is bound to ask you which cells contain which structures.

Why do we need to know about the structure of cells? Because biological structure is often closely related to function. (Watch out for this connection: it's a favorite theme for the AP Biology Exam.) And, more importantly, because the testing board likes to test you on it!

Structural Characteristics of Different Cell Types			
Structure	Prokaryote	Plant Cell	Animal Cell
Cell Wall	Yes	Yes	No
Plasma Membrane	Yes	Yes	Yes
Organelles	No	Yes	Yes
Nucleus	No	Yes	Yes
Centrioles	No	No	Yes
Ribosomes	Yes	Yes	Yes

Transport: The Traffic Across Membranes

We've talked about the structure of cell membranes; now let's discuss how molecules and fluids pass through the plasma membrane. What are some of the patterns of membrane transport? The ability of molecules to move across the cell membrane depends on two things: (1) the semipermeability of the plasma membrane and (2) the size and charge of particles that want to get through.

Since the plasma membrane is composed primarily of phospholipids, lipid-soluble substances cross the membrane without any resistance. Why? Because "like dissolves like." The lipid bilayer is a hydrophobic sandwich (hydrophilic outside and a hydrophobic interior), and only hydrophobic things can pass that zone. If a substance is hydrophilic, the bilayer won't let it pass without assistance, called **facilitated transport**.

Facilitated transport depends upon a number of proteins that act as tunnels through the membrane. Channels are very specialized types of tunnels that let only certain things through.

The most famous of these are **aquaporins**, which are water-specific channels. Although water is polar, there are typically sufficient aquaporins for water to traverse the membrane whenever it wishes without forming traffic jams. However, without aquaporins, no water would be able to cross the membrane.

Passive Transport: Simple and Facilitated Diffusion

Now we know how things cross the membrane, but let's talk about why they cross the membrane. The simple answer is "To get to the other side." If there is a high concentration of something in one area, it will move to spread out and diffuse into an area with a lower concentration, even if that means entering or exiting a cell. In other words, the substance moves *down* a concentration gradient. This is called diffusion. When the molecule that is diffusing is hydrophobic, the diffusion is called **simple diffusion** because the small nonpolar molecule can just drift right through the membrane without trouble. When the diffusion requires the help of a channel-type protein, it is called **facilitated diffusion**. Anytime that a substance is moving by diffusion, it is called **passive transport** because there is no outside energy required to cause the movement. It's like riding a bicycle downhill. The molecules are just "going with the flow."

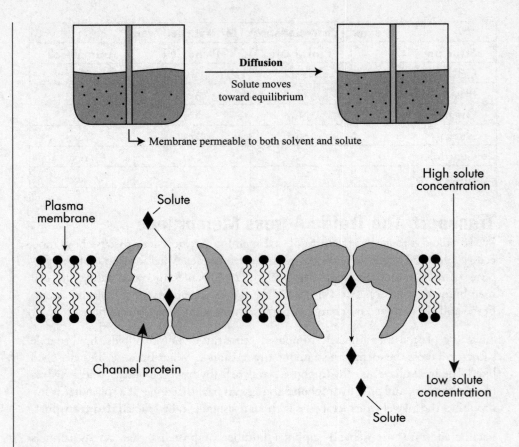

Membrane permeable to both solvent and solute

Osmosis

When the molecule that is diffusing is water, the process is called **osmosis**. Water always wants to move from an area where it is most concentrated to an area where it is least concentrated. Water is not usually pure water, but since it is so great at dissolving things, it usually exists as a solution. A solution is made when a liquid solvent dissolves solute particles. So, if we rethink water moving down its concentration gradient, this is the same as saying that it wants to go from where there is less solute (ie. watery solution) to where there is more solute (ie. concentrated solution). It's as if the water is moving to dilute the concentrated solute particles.

Often, situations discussing osmosis set it up so that there are two areas and the water can flow freely, but the solute cannot. For example, if a chamber containing water and a chamber containing a sucrose solution are connected by a semipermeable membrane that allows water but not sucrose to cross, diffusion of sucrose between the chambers cannot occur. In this case, osmosis draws water into the sucrose chamber to reduce the sucrose concentration. This will reduce the total volume of the water chamber. Water will flow into the sucrose chamber until the concentration is the same across the membrane.

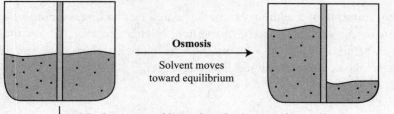

Osmosis

Solvent moves
toward equilibrium

↳ Membrane permeable to solvent but impermeable to solute

In both diffusion and osmosis, the final result is that solute concentrations are the same on both sides of the membrane. The only difference is that in diffusion the membrane is usually permeable to solute, and in osmosis it is not.

The term **tonicity** is used to describe osmotic gradients. If the environment is **isotonic** to the cell, the solute concentration is the same inside and outside. A **hypertonic** solution has more total dissolved solutes than the cell, while a **hypotonic** solution has less. The tendency of water to move down its concentration gradient (into cells) can be a powerful force; it can cause cells to explode and can overcome gravity. You may also hear the terms *isosmotic*, *hyperosmotic*, and *hypoosmotic*. These words are usually used when comparing two solutions, while tonicity terms are used when comparing a solution to a cell. Other than this, the definitions are similar; isotonic and isosmotic both describe situations in which the concentration is the same on either side of a membrane, for example. The only difference is what type of membrane.

Water potential (Ψ) is the measure of potential energy in water and describes the eagerness of water to flow from an area of high water potential to an area of low water potential. It is affected by two factors: pressure potential (Ψ_p) and solute potential (Ψ_s). The equation for **solute (or osmotic) potential** is listed on the AP Biology Equations and Formulas sheet and describes the effect of solute concentration on water flow. Adding a solute lowers the water potential of a solution, causing water to be less likely to leave this solution and more likely to flow into it. This makes sense because the added solute makes the solution more concentrated, and the water is now unlikely to diffuse away. The more solute molecules present, the more negative the solute potential is.

A red blood cell dropped into a hypotonic solution (such as distilled water) will expand because water will move into the cell, an area of lower water potential. Eventually, the red blood cell will pop. If a similar experiment is done with a plant cell, water will still move into the cell, but the cell wall will exert pressure, increasing the water potential and limiting the gain of water. Water potential also explains how water moves from soil into plant roots and how plants transport water from roots to leaves to support photosynthesis.

Remember It Like This
An area of high water potential is an area of high water concentration (e.g., low solute).

The concentration of a solution can be calculated by dividing the number of moles of solute by the volume (in liters) of solution. Highly concentrated (or stock) solutions can be diluted to make less concentrated solutions. The equation to do this is on the AP Biology Equations and Formulas sheet.

$$C_i V_i = C_f V_f$$

This equation tells you how much (V_i) of a concentrated solution (at C_i) is required to make a more dilute solution (C_f) of a certain volume (V_f). The amount of solvent added will be the final volume of the solution (V_f) minus whatever amount of stock solution you need to add. For example, if you have a bottle of 2 M solution of Tris buffer and would like to make 50 mL of 0.1 M Tris.

$$C_i V_i = C_f V_f$$
$$(2\ M)(V_i) = (0.1\ M)(0.050\ L)$$
$$V_i = 0.0025\ L,\ or\ 2.5\ mL$$

Thus, you would add 2.5 mL of the concentrated (or stock) Tris solution to a new tube or bottle and then add water (the solvent) up to a final volume of 50 mL. In other words, you should add 47.5 mL of water.

Active Transport

Suppose a substance wants to move in the opposite direction—from a region of lower concentration to a region of higher concentration. A transport protein can help usher the substance across the plasma membrane, but it's going to need energy to accomplish this. This time it's like riding a bicycle uphill. Compared with riding downhill, riding uphill takes a lot more work. Movement against the natural flow is called **active transport**.

But where does the protein get this energy? Some proteins in the plasma membrane are powered by ATP. The best example of active transport is a special protein called the **sodium-potassium pump**. It ushers out three sodium ions (Na^+) and brings in two potassium ions (K^+) across the cell membrane. This pump depends on ATP to get ions across that would otherwise remain in regions of higher concentration. Primary active transport occurs when ATP is directly utilized to transport something. Secondary active transport occurs when something is actively transported using the energy captured from the movement of another substance flowing down its concentration gradient.

We've now seen that small substances can cross the cell membrane by:

- simple diffusion
- facilitated transport
- active transport

Endocytosis

When the particles that want to enter a cell are just too large, the cell uses a portion of the cell membrane to engulf the substance. The cell membrane forms a pocket, pinches in, and eventually forms either a vacuole or a vesicle. This process is called **endocytosis**.

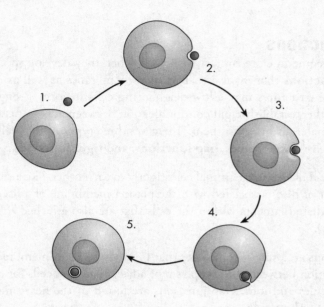

Three types of endocytosis exist: **pinocytosis**, **phagocytosis**, and **receptor-mediated endocytosis**. In pinocytosis, the cell ingests liquids ("cell-drinking"). In phagocytosis, the cell takes in solids ("cell-eating"). A special type of endocytosis, receptor-mediated endocytosis, involves cell surface receptors that work in tandem with endocytic pits that are lined with a protein called clathrin. When a particle, or ligand, binds to one of these receptors, the ligand is brought into the cell by the invagination or "folding in" of the cell membrane. A vesicle then forms around the incoming ligand and carries it into the cell's interior.

Bulk Flow

Other substances move by **bulk flow**. Bulk flow is the one-way movement of fluids brought about by *pressure*. For instance, the movement of blood through a blood vessel and the movement of fluids in xylem and phloem of plants are examples of bulk flow.

Dialysis

Dialysis is the diffusion of *solutes* across a selectively permeable membrane. Special membranes that have holes of a certain size within them can be used to sort substances by using the process of diffusion. Kidney dialysis is a specialized process by which the blood is filtered using machines and concentration gradients. Things present at high levels will naturally diffuse out of the blood; dialysis just gives them the opportunity.

Exocytosis

Sometimes large particles are transported *out* of the cell. In **exocytosis**, a cell ejects waste products or specific secretion products, such as hormones, by the fusion of a vesicle with the plasma membrane, which then expels the contents into the extracellular space. Think of it as reverse endocytosis.

Cell Junctions

When cells come in close contact with each other, they develop specialized **intercellular junctions** that involve their plasma membranes as well as other components. These structures may allow neighboring cells to form strong connections with each other, establish rapid communication between adjacent cells, or prevent passage of materials between them. There are three types of intercellular contacts in animal cells: **desmosomes**, **gap junctions**, and **tight junctions**.

Desmosomes hold adjacent animal cells tightly to each other, like a rivet. They consist of a pair of discs associated with the plasma membrane of adjacent cells, plus the intermediate filaments within the cells that are also attached to the discs (see figure below).

Gap junctions are protein complexes that form channels in membranes and allow communication between the cytoplasm of adjacent animal cells for the transfer of small molecules and ions. Gap junctions are found in the heart and in compact bone and smooth muscle.

Tight junctions are tight connections between the membranes of adjacent animal cells. They're so tight that there is no space between the cells. Cells connected by tight junctions seal off body cavities and prevent leaks. Tight junctions are found in the small intestine.

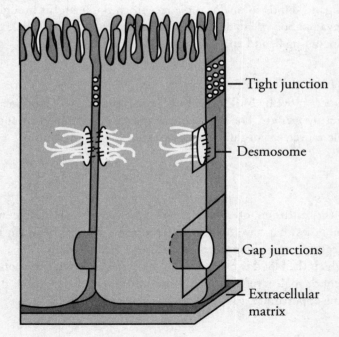

—Tight junction

—Desmosome

—Gap junctions

—Extracellular matrix

Cell Communication

Communication is important for organisms and involves transduction of stimulatory or inhibitory signals from other cells, organisms, or the environment. Unicellular organisms detect and respond to environmental signals. For example, they can use cell communication to locate a suitable mate, adapt metabolism rates in response to changes in nutrient availability, or make their numbers known to other members of their species (a phenomenon known as **quorum sensing**).

Taxis is the movement of an organism in response to a stimulus and can be positive (toward the stimulus) or negative (away from the stimulus). Taxes are innate behavioral responses, or instincts. **Chemotaxis** is movement in response to chemicals. For example, bacteria can control flagella rotation to direct their motion, thus avoiding repellents (such as poisons), or helping them find favorable locations with high concentrations of attractants (such as food). In our bodies, neutrophils use chemotaxis to respond to an infection and are the first responders to inflammation.

The cells of multicelled organisms must communicate with one another to coordinate the activities of the organism as a whole. Signaling can be short-range (affecting only nearby cells) or long-range (affecting cells throughout the organism). It can be done by cell junctions or signaling molecules called **ligands** that bind to **receptors** and trigger a response. Hormones and neurotransmitters are common signaling molecules, and they will be covered in Chapter 11.

Signal transduction is the process by which an external signal is transmitted to the inside of a cell. It usually involves the following three steps:

1. a signaling molecule binding to a specific receptor
2. activation of a signal transduction pathway
3. production of a cellular response

Hydrophobic signaling molecules, like cholesterol-based steroid hormones, can diffuse across the plasma membrane. These molecules can bind receptors inside the cell and often travel to the nucleus where they regulate gene expression.

For signaling molecules that cannot enter the cell, a plasma membrane receptor is required. Plasma membrane receptors form an important class of integral membrane proteins that transmit signals from the extracellular space into the cytoplasm. Each receptor binds a particular molecule in a highly specific way. There are three classes of membrane receptors.

1. **Ligand-gated ion channels** in the plasma membrane open an ion channel upon binding a particular ligand. An example is the ligand-gated sodium channel on the surface of a skeletal muscle cell at the neuromuscular junction. This channel opens in response to acetylcholine, and a massive influx of sodium depolarizes the muscle cell and causes it to contract.

2. **Catalytic** (or **enzyme-linked**) **receptors** have an enzymatic active site on the cytoplasmic side of the membrane. Enzyme activity is initiated by ligand binding at the extracellular surface. The insulin receptor is an example of an enzyme-linked receptor. After binding insulin, enzymatic activity initiates a complex signaling pathway that allows the cell to grow, synthesize lipids, and import glucose.

3. A **G-protein-linked receptor** does not act as an enzyme, but instead will bind a different version of a G-protein (often GTP or GDP) on the intracellular side when a ligand is bound extracellularly. This causes activation of secondary messengers within the cell. One important second messenger is cyclic AMP (cAMP). It is known as a "universal hunger signal" because it is the second messenger of the hormones epinephrine and glucagon, which cause energy mobilization. Secondary messengers are usually small molecules that can move quickly through the cell. They can be made and destroyed quickly and help the signal amplify throughout the cell.

Signal transduction in eukaryotic cells usually involves many steps and complex regulation. In contrast, bacterial cells usually use a simpler two-component regulatory system in transduction pathways.

CELL COMMUNICATION IN PLANTS

Plants do not have a nervous system but can produce several proteins found in animal nervous systems, such as certain neurotransmitter receptors. Plants can also generate electrical signals in response to environmental stimuli, and this can affect flowering, respiration, photosynthesis, and wound healing. Light receptors are common in plants and help link environmental cues to biological processes such as seed germination, the timing of flowering, and chlorophyll production. Some plants can also use chemicals to communicate with nearby plants. For example, wounded tomatoes produce a volatile chemical as an alarm signal. This warns nearby plants and allows them to prepare a defense or immune response.

KEY TERMS

cells
light microscopes
electron microscopes
prokaryotic cells
eukaryotic cells
cytoplasm
nucleoid
cell wall
plasma membrane
flagella
capsule
organelles
phosholipid bilayer
peripheral proteins
integral proteins
transmembrane proteins
fluid-mosaic model
adhesion proteins
receptor proteins
transport proteins
channel proteins
cell surface markers
carbohydrate side chains
nucleus
chromosomes
nucleolus
ribosomes
endoplasmic reticulum (ER)
rough ER
smooth ER
Golgi bodies
vesicles
mitochondria
adenosine triphosphate (ATP)
lysosomes
centrioles
microtubule organizing centers
  (MTOCs)
vacuoles
peroxisomes
cytoskeleton
microtubules
microfilaments

tubulin
cilia
Euglena
Paramecium
cell wall
chitin
chloroplasts
cell sap
facilitated transport
aquaporins
simple diffusion
facilitated diffusion
passive transport
osmosis
tonicity
isotonic
hypertonic
hypotonic
water potential
solute (or osmotic) potential
solutes
active transport
sodium-potassium pump
endocytosis
pinocytosis
phagocytosis
receptor-mediated endocytosis
bulk flow
dialysis
exocytosis
intercellular junctions
desmosomes
gap junctions
quorum sensing
tight junctions
taxis
chemotaxis
ligands
receptors
signal transduction
ligand-gated ion channel
catalytic (or enzyme-linked) receptor
G-protein-linked receptor

Summary

- Living things are composed of cells, the smallest unit of living organisms. Cells may be categorized into prokaryotes, which do not have a nucleus or membrane bound organelles, and eukaryotes, which have a nucleus and membrane-bound organelles. Features of prokaryotic cells (like bacteria, for example) can include the following:
  - plasma membrane
  - flagella and cilia
  - cell wall
  - nucleoid region
  - ribosomes
  - circular chromosome
  - cytoplasm

- Components of the eukaryotic cells include the following:
  - nucleus
  - nucleolus
  - rough endoplasmic reticulum
  - smooth endoplasmic reticulum
  - Golgi body
  - ribosomes
  - mitochondria
  - vacuole
  - lysosome
  - peroxisomes
  - centrioles (animals only)
  - cytoplasm
  - plasma membrane
  - cytoskeleton (microtubules, microfilaments)
  - central vacuole (plants only)
  - cell wall (plants only)

- The plasma membrane is composed of a phospholipid bilayer. The hydrophobic nature of the inside of the membrane is responsible for its selective permeability. Transport through the membrane is dependent on size and polarity of the molecules and concentration gradients. There are a few different types of cellular transport:
  - simple diffusion
  - facilitated transport
  - active transport
  - bulk flow
  - dialysis
  - exocytosis
  - endocytosis (pinocytosis, phagocytosis, receptor-mediated endocytosis)

- Cells are connected by tight junctions, desmosomes, and gap junctions, which all function to connect adjacent cells.

- Sometimes cells do not need to pass any molecules through the membrane, but instead they just pass a message. This is signal transduction. A ligand binds to a receptor on the outside of the cell and causes changes to the inside of the cell. Ligand-gated ion channels, catalytic receptors, and G-protein-linked receptors are common examples.

Chapter 5 Drill

Answers and explanations can be found in Chapter 16.

1. Movement of substances into the cell is largely dependent on the size, polarity, and concentration gradient of the substance. Which of the following represents an example of active transport of a substance into a cell?

   (A) Diffusion of oxygen into erythrocytes (red blood cells) in the alveolar capillaries of the lungs
   (B) Influx of sodium ions through a voltage-gated ion channel in a neuron cell during an action potential
   (C) The sodium-potassium pump, which restores resting membrane potentials in neurons through the use of ATP
   (D) Osmosis of water into an epithelial cell lining the lumen of the small intestine

2. The development of electron microscopy has provided key insights into many aspects of cellular structure and function, which had previously been too small to be seen. All of the following would require the use of electron microscopy for visualization EXCEPT

   (A) the structure of a bacteriophage
   (B) the matrix structure of a mitochondrion
   (C) the shape and arrangement of bacterial cells
   (D) the pores on the nuclear membrane

3. *Vibrio cholerae* (shown below) are highly pathogenic bacteria that are associated with severe gastrointestinal illness and are the causative agent of cholera. In extreme cases, antibiotics are prescribed that target bacterial structures that are absent in animal cells. Which of the following structures is most likely targeted by antibiotic treatment?

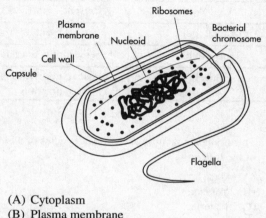

   (A) Cytoplasm
   (B) Plasma membrane
   (C) Ribosomes
   (D) Cell wall

Question 4 is based on the information and table below.

A new unicellular organism has recently been identified living in thermal pools in Yellowstone National Park. The thermal pools have average temperatures of 45°C, a pH of approximately 3.2, and high concentrations of sulfur-containing compounds. To identify the organism, a microbiologist performs a series of tests to evaluate its structural organization. The table below summarizes the microscopy data of the newly identified organism.

Cellular Structure	Analysis
Plasma Membrane	Present
Cell Wall	Present, very thick
Mitochondria	Absent
Ribosomes	Present, highly abundant
Flagella	Present, peritrichous organization

4. This organism is most likely a new species of which of the following?

    (A) Algae
    (B) Protozoa
    (C) Bacteria
    (D) Fungi

Question 5 represents a question requiring a numeric answer.

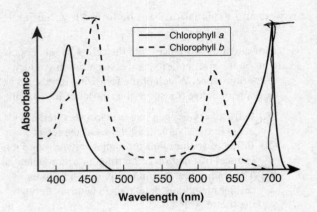

5. Chlorophyll is a green pigment present in the chloroplasts of algae and plants. It is essential for catalyzing the light-dependent cycles in photosynthesis. A scientist purifies both forms of chlorophyll (*a* and *b*) from plant chloroplasts and evaluates them for light absorption using a spectrophotometer. Using the spectrophotometer data provided above, at what wavelength is the absorbance of chlorophyll *a* at its maximum? Give your answer to the nearest whole number.

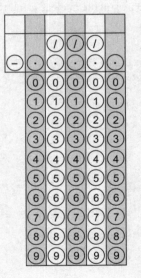

6. In a eukaryotic cell, which of the following organelles directly work together?

   (A) Nuclear envelope, nucleolus, vacuoles, centrioles
   (B) Ribosomes, rough endoplasmic reticulum, Golgi bodies, plasma membrane
   (C) Mitochondria, ribosomes, lysosomes, chloroplasts
   (D) Centrioles, nucleolus, smooth endoplasmic reticulum, lysosomes

7. A lipid-soluble hormone, estrogen, is secreted from the ovaries. This molecule travels through the body via the bloodstream. A researcher was interested in reducing estrogen's effect in order to determine the response of decreased estrogen on the organism. Which of the following is a valid strategy for reducing effects of estrogen on the whole research organism?

   (A) Treat with a competitive inhibitor drug that blocks all receptors at the plasma membrane
   (B) Treat with lipid-soluble testosterone
   (C) Treat with a lipid-soluble noncompetitive inhibitor that specifically reduces estrogen binding to the intracellular receptor
   (D) Remove ovaries of the organism

8. Which type of cell would be the most useful for studying the rough endoplasmic reticulum?

   (A) Neurons firing action potentials
   (B) Insulin-making cells in the pancreas
   (C) Bacterial cells in a colony
   (D) White blood cells

9. *Paramecium* is a single-celled protist that lives in freshwater habitats. In these conditions, *Paramecium* has evolved strategies to handle the potential consequences of inhabiting this hypotonic environment. One of these strategies could be

   (A) contractile vacuoles, which expel water forcefully
   (B) increased aquaporins in its cellular membrane
   (C) many cilia covering its surface
   (D) salt receptors on its surface to seek out less concentrated areas

10. A co-transporter is something that moves two substances across a membrane, one passively and the other actively. The Na^+K^+ ATPase transports sodium and potassium ions across the plasma membrane against their concentration gradients. This pump is not considered a co-transporter because

   (A) ATP is produced through this transporter
   (B) both ions are moved via active transport
   (C) both ions are moved via passive transport
   (D) ATP hydrolysis does not occur during transport

REFLECT

Respond to the following questions:

- Which topics from this chapter do you feel you have mastered?

- Which content topics from this chapter do you feel you need to study more before you can answer multiple-choice questions correctly?

- Which content topics from this chapter do you feel you need to study more before you can effectively compose a free response?

- Was there any content that you need to ask your teacher or another person about?

Chapter 6
Cellular Energetics

BIOENERGETICS

In Chapter 4, we discussed some of the more important organic molecules. But what makes these molecules so important? Glucose, starch, and fat are all energy-rich. However, the energy is packed in the chemical bonds holding the molecules together. To carry out the processes necessary for life, cells must find a way to release the energy in these bonds when they need it and store it away when they don't. The study of how cells accomplish this is called **bioenergetics**. Generally, bioenergetics is the study of how energy from the sun is transformed into energy in living things.

During chemical reactions, such bonds are either broken or formed. This process involves energy, no matter which direction we go. Every chemical reaction involves a change in energy.

ADENOSINE TRIPHOSPHATE (ATP)

We've all heard the expression *"nothing in life is free."* The same holds true in nature.

> Energy cannot be created or destroyed. In other words, the sum of energy in the universe is constant.

This rule is called the **First Law of Thermodynamics**. As a result, the cell cannot take energy out of thin air. Rather, it must harvest it from somewhere.

The **Second Law of Thermodynamics** states that energy transfer leads to less organization. That means the universe tends toward disorder (or **entropy**).

TYPES OF REACTIONS

Exergonic reactions are those in which the products have *less* energy than the reactants. Simply put, energy is given off during the reaction.

Let's look at an example. The course of a reaction can be represented by an **energy diagram**. Here's an energy diagram for an exergonic reaction.

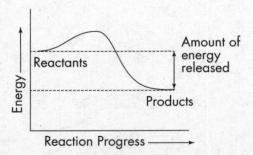

You'll notice that energy is represented along the y-axis. Based on the diagram, we see our reaction released energy. An example of an exergonic reaction is the oxidizing of molecules in mitochondria of cells, which then releases the energy stored in the chemical bonds.

Reactions that require an input of energy are called **endergonic reactions**. You'll notice that the products have *more* energy than the reactants. An example is plants' use of carbon dioxide and water to form sugars.

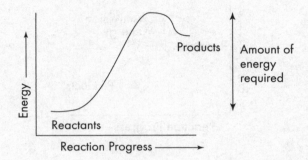

GIBBS FREE ENERGY

A practical way to discuss thermodynamics is the mathematical notion of **Gibbs free energy**.

$$\Delta G = \Delta H - T\Delta S$$

This equation is on the AP Biology Equations and Formulas sheet. T denotes temperature, H denotes **enthalpy** (the measure of energy in a thermodynamic system), and S denotes entropy. The change in the Gibbs free energy of a reaction determines whether the reaction is favorable (spontaneous, negative ΔG) or unfavorable (nonspontaneous, positive ΔG). In other words, this equation can be used to figure out if (without adding extra energy) the reactants will stay as they are or be converted to products.

Spontaneous reactions, ones that occur without a net addition of energy, have $\Delta G < 0$. They occur with energy to spare. Reactions with a negative ΔG are exergonic (energy exits the system); reactions with a positive ΔG are endergonic. Endergonic reactions occur only if energy is added. In the lab, energy is added in the form of heat; in the body, endergonic reactions are driven by reaction coupling to exergonic or favorable reactions.

Activation Energy

Even though exergonic reactions release energy, the reaction might not occur naturally without a little bit of energy to get things going. This is because the reactants must first turn into an intermediate state, called the **transition state**, before turning into the products. The transition state is sort of a reactants-products hybrid state that is difficult to achieve. In order to reach this transition state, a certain amount of energy is required. This is called the **activation energy**.

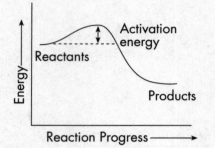

You'll notice that we needed a little energy to get us going. That's because chemical bonds must be broken before new bonds can form. This energy barrier—the hump in the graph—is called the activation energy. Once a set of reactants has reached its activation energy, the rest of the reaction is all downhill. Reaching the transition state is the tough part.

Enzymes

A catalyst is something that speeds something up. **Enzymes** are biological catalysts that speed up reactions. They accomplish this by lowering the activation energy and helping the transition state to form. Thus, the reaction can occur more quickly because the tricky transition state is not as much of a hurdle to overcome. Enzymes do NOT change the energy of the starting point or the ending point of the reaction. They only lower the activation energy.

Enzyme Specificity

Most of the crucial reactions that occur in the cell require enzymes. Yet enzymes themselves are highly specific—in fact, each enzyme catalyzes only one kind of reaction. This is known as **enzyme specificity**. Because of this, enzymes are usually named after the molecules they target. In enzymatic reactions, the targeted molecules are known as **substrates**. For example, maltose, a disaccharide, can be broken down into two glucose molecules. Our substrate, maltose, gives its name to the enzyme that catalyzes this reaction: maltase.

Many enzymes are named simply by replacing the suffix of the substrate with –*ase*. Using this nomenclature, malt*ose* becomes malt*ase*.

Enzyme-Substrate Complex

Enzymes have a unique way of helping reactions along. As we just saw, the reactants in an enzyme-assisted reaction are known as substrates. During a reaction, the enzyme's job is to bring the transition state about by helping the substrate(s) get into position. It accomplishes this through a special region on the enzyme known as an **active site**.

The enzyme temporarily binds one or more of the substrates to its active site and forms an **enzyme-substrate complex**. Let's take a look:

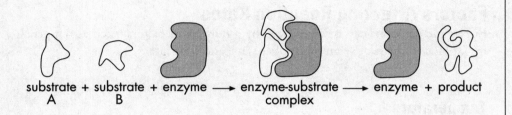

substrate + substrate + enzyme ⟶ enzyme-substrate ⟶ enzyme + product
A   B   complex

Once the reaction has occurred and the product is formed, the enzyme is released from the complex and restored to its original state. Now the enzyme is free to react again with another bunch of substrates.

By binding and releasing substrates over and over again, the enzyme speeds the reaction along, enabling the cell to release much-needed energy from various molecules. Here is a quick review on the function of enzymes.

Enzymes Do:

- increase the rate of a reaction by lowering the reaction's activation energy
- form temporary enzyme-substrate complexes
- remain unaffected by the reaction

Enzymes Don't:

- change the reaction
- make reactions occur that would otherwise not occur at all

Induced Fit

However, scientists have discovered that enzymes and substrates don't fit together quite so seamlessly. It appears that the enzyme has to change its shape slightly to accommodate the shape of the substrates. This is called **induced fit**. Sometimes certain factors are involved in activating the enzyme and making it capable of binding the substrate. Because the fit between the enzyme and the substrate must be perfect, enzymes operate only under a strict set of biological conditions.

Enzymes Don't Always Work Alone

Enzymes sometimes need a little help in catalyzing a reaction. Those factors are known as **cofactors**. Cofactors can be either organic molecules called **coenzymes** or inorganic molecules or ions. The inorganic cofactors are usually metal ions (Fe^{2+}, Mg^{2+}). Vitamins are examples of organic coenzymes. Your daily dose of vitamins is important for just this reason: the vitamins are active and necessary participants in crucial chemical reactions.

Factors Affecting Reaction Rates

Enzymatic reactions can be influenced by a number of factors, such as temperature, pH, and the relative amounts of enzyme and substrate.

Temperature

Break Down the Words
To de-nature is to un-nature something. To change its nature. To remove the characteristics that make it unique. Without the proper special folding, a protein loses its nature.

The rate of a reaction increases with increasing temperature, up to a point, because an increase in the temperature of a reaction increases the chance of collisions among the molecules. But too much heat can damage an enzyme. If a reaction is conducted at an excessively high temperature (above 42°C), the enzyme loses its three-dimensional shape and becomes inactivated. Enzymes damaged by heat and deprived of their ability to catalyze reactions are said to be **denatured**.

Here's one thing to remember: all enzymes operate at an ideal temperature. For most human enzymes, this temperature is body temperature, 37°C.

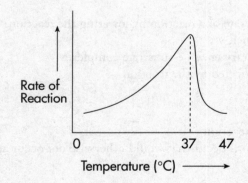

Not all organisms have a constant body temperature. For example, ectotherms depend on the environment to control their varying body temperatures. Q_{10} is a measure of temperature sensitivity of a physiological process or enzymatic reaction rate. The equation for Q_{10} is included on the AP Biology Equations and Formulas sheet.

$$Q_{10} = \left(\frac{k_2}{k_1} \right)^{\frac{10}{T_2 - T_1}}$$

The temperature unit must be in either Celsius or Kelvin, and the same unit must be used for T_1 and T_2. The two reaction rates (k_1 and k_2) must also have the same unit. Q_{10} has no unit; it is the factor by which the rate of a reaction increases due to a temperature increase. The more temperature-dependent a reaction is, the higher Q_{10} will be. Reactions with $Q_{10} = 1$ are temperature independent.

pH

Enzymes also function best at a particular pH. For most enzymes, the optimal pH is at or near a pH of 7.

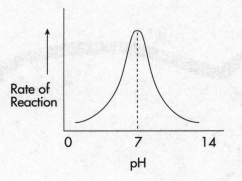

Other enzymes operate at a low pH. For instance, pepsin, the digestive enzyme found in the stomach, is most effective at an extremely acidic pH of 2.

Enzyme Regulation

We know that enzymes control the rates of chemical reactions. But what regulates the activity of enzymes? It turns out that a cell can control enzymatic activity by regulating the conditions that influence the shape of the enzyme. Enzymes can be turned on/off by things that bind to them. Sometimes these things can bind at the active site, and sometimes they bind at other sites, called **allosteric sites**.

If the substance has a shape that fits the active site of an enzyme (i.e., similar to the substrate), it can compete with the substrate and block the substrate from getting into the active site. This is called **competitive inhibition**. If there was enough substrate available, the substrate would out-compete the inhibitor and the reaction would occur. You can always tell a competitive inhibitor based on what happens when you flood the system with lots of substrate.

If the inhibitor binds to an allosteric site, it is an **allosteric inhibitor** and it is **noncompetitive inhibition**. A noncompetitive inhibitor generally distorts the enzyme shape so that it cannot function. The substrate can still bind, but the enzyme will not be able to catalyze the reaction.

REACTION COUPLING

As we just saw, almost everything an organism does requires energy. How, then, can the cell acquire the energy it needs without becoming a major mess? Fortunately, it's through adenosine triphosphate (ATP).

ATP, as the name indicates, consists of a molecule of adenosine bonded to three phosphates. The great thing about ATP is that an enormous amount of energy is packed into those phosphate bonds.

The body can perform difficult endergonic reactions by coupling them to ATP hydrolysis, which is very exergonic.

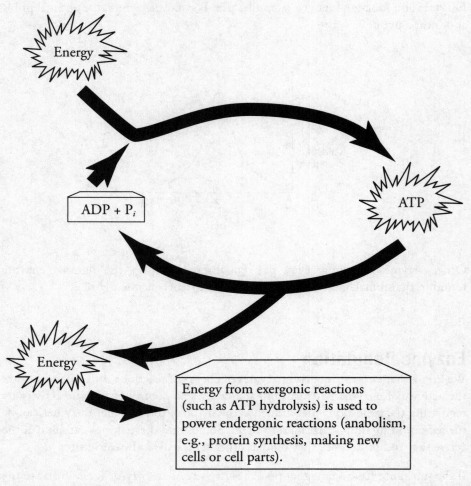

Energy from exergonic reactions (such as ATP hydrolysis) is used to power endergonic reactions (anabolism, e.g., protein synthesis, making new cells or cell parts).

When a cell needs energy, it takes one of these potential-packed molecules of ATP and splits off the third phosphate, forming adenosine diphosphate (ADP) and one loose phosphate (P_i), while releasing energy in the process.

$$ATP \rightarrow ADP + P_i + energy$$

The energy released from this reaction can then be put to whatever use the cell pleases. Of course, this doesn't mean that the cell is above the laws of thermodynamics. But within those constraints, ATP is the best source of energy the cell has available. It is relatively neat (only one bond needs to be broken to release that energy) and relatively easy to form. Organisms can use exergonic processes that increase energy, like breaking down ATP, to power endergonic reactions, like building organic macromolecules.

Sources of ATP

But where does all this ATP come from? It can be formed in several ways, but the bulk of it comes from a process called **cellular respiration**. Cellular respiration is a process of breaking down sugar and making ATP.

In autotrophs, the sugar is made during **photosynthesis**. In heterotrophs, the glucose comes from the food we eat. We will start by going over photosynthesis, since in plants it is a precursor to cellular respiration.

PHOTOSYNTHESIS

Plants and algae are producers. All they do is bask in the sun, churning out the glucose necessary for life.

As we discussed earlier, photosynthesis is the process by which light energy is converted to chemical energy. Here's an overview of photosynthesis.

$$6CO_2 + 6H_2O \longrightarrow C_6H_{12}O_6 + 6O_2$$

You'll notice from this equation that carbon dioxide and water are the raw materials plants use to manufacture simple sugars. But remember, there's much more to photosynthesis than what's shown in the simple reaction above.

There are two stages in photosynthesis: the **light reactions** (also called the light-dependent reactions) and the **dark reactions** (also called the light-independent reactions). The whole process begins when the **photons** (or "energy units") of sunlight strike a leaf, activating chlorophyll and exciting electrons. The activated chlorophyll molecule then passes these excited electrons down to a series of electron carriers, ultimately producing ATP and NADPH. Both of these products, along with carbon dioxide, are then used in the dark reactions (light-independent) to make carbohydrates. Along the way, water is also split and oxygen gets released.

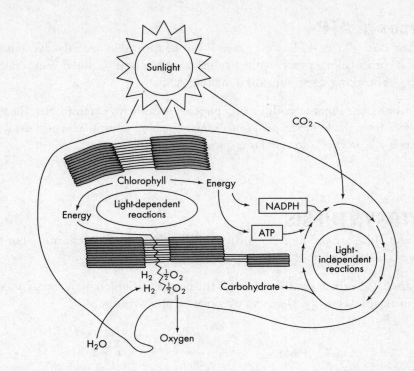

We'll soon see that this beautifully orchestrated process occurs thanks to a whole host of special enzymes and pigments. But before we turn to the stages in photosynthesis, let's talk about where photosynthesis occurs.

Chloroplast Structure

The leaves of plants contain lots of chloroplasts, which are the primary sites of photosynthesis.

Now let's look at an individual chloroplast. If you split the membranes of a chloroplast, you'll find a fluid-filled region called the **stroma**. Inside the stroma are structures that look like stacks of coins. These structures are the **grana**.

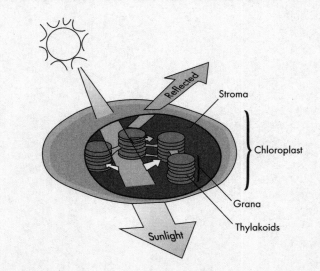

The many disk-like structures that make up the grana are called the **thylakoids.** They contain chlorophyll, the light-absorbing pigment that drives photosynthesis, as well as enzymes involved in the process.

Many light-absorbing pigments participate in photosynthesis. Some of the more important ones are **chlorophyll *a*, chlorophyll *b*,** and **carotenoids**. These pigments are clustered in the thylakoid membrane into units called antenna complexes.

All of the pigments within a unit are able to "gather" light, but they're not able to "excite" the electrons. Only one special molecule—located in the **reaction center**— is capable of transforming light energy to chemical energy. In other words, the other pigments, called **antenna pigments**, "gather" light and "bounce" the energy to the reaction center.

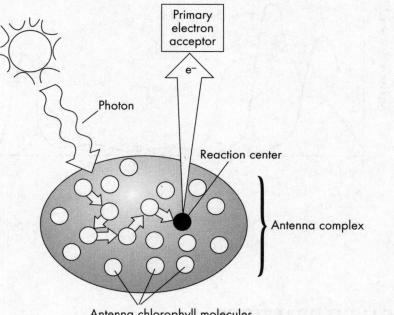

There are two types of reaction centers: **photosystem I (PS I)** and **photosystem II (PS II)**. The principal difference between the two is that each reaction center has a specific type of chlorophyll—chlorophyll *a*—that absorbs a particular wavelength of light. For example, **P680**, the reaction center of photosystem II, has a maximum absorption at a wavelength of 680 nanometers. The reaction center for photosystem I—**P700**—best absorbs light at a wavelength of 700 nanometers.

When light energy is used to make ATP, it is called **photophosphorylation**. Autotrophs are using light (that's *photo*) and ADP and phosphates (that's *phosphorylation*) to produce ATP.

An **absorption spectrum** shows how well a certain pigment absorbs electromagnetic radiation. Light absorbed is plotted as a function of radiation wavelength. This spectrum is the opposite of an **emission spectrum**, which gives information on which wavelengths are emitted by a pigment. The absorption spectrum for chlorophyll *a*, chlorophyll *b*, and carotenoids is below. You can see that chlorophyll *a* and chlorophyll *b* absorb blue and red light quite well but do not absorb light in the green part of the spectrum. Light in the green range of wavelengths is reflected, and this is why chlorophyll and many plants are green. Carotenoids absorb light on the blue-green end of the spectrum, but not on the other end. This is why plants rich in carotenoids are yellow, orange, or red.

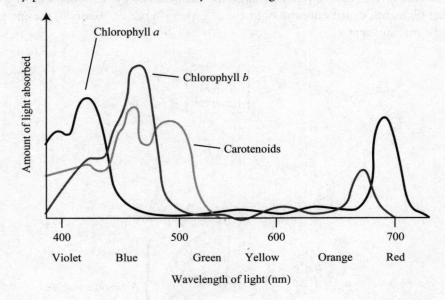

THE LIGHT REACTIONS

When a leaf captures sunlight, the energy is sent to P680, the reaction center for photosystem II. The activated electrons are trapped by P680 and passed to a molecule called the primary acceptor, and then they are passed down to carriers in the electron transport chain.

To replenish the electrons in the thylakoid, water is split into oxygen, hydrogen ions, and electrons. That process is called **photolysis**. The electrons from photolysis replace the missing electrons in photosystem II.

As the energized electrons from photosystem II travel down the electron transport chain, they pump hydrogen ions into the thylakoid lumen. A proton gradient is established. As the hydrogen ions move back into the stroma through ATP synthase, ATP is produced.

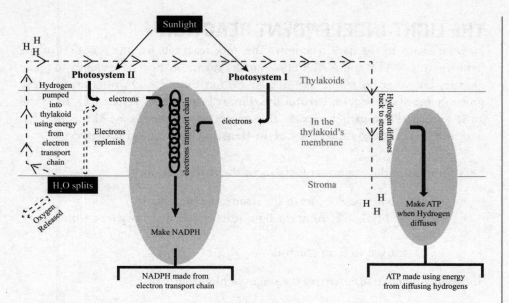

After the electrons leave photosystem II, they go to photosystem I. The electrons are passed through a second electron transport chain until they reach the final electron acceptor NADP⁺ to make **NADPH**.

> The light reaction occurs in the thylakoids.

- P680 in photosystem II captures light and passes excited electrons down an electron transport chain to produce ATP.
- P700 in photosystem I captures light and passes excited electrons down an electron transport chain to produce NADPH.
- A molecule of water is split by sunlight, releasing electrons, hydrogen, and free O_2.

The light-dependent reactions occur in the grana of chloroplasts, where the thylakoids are found. Remember: the light-absorbing pigments and enzymes for the light-dependent reactions are found within the thylakoids.

THE LIGHT-INDEPENDENT REACTION

Now let's turn to the dark reactions. The dark reactions use the products of the light reaction—ATP and NADPH—to make sugar. We now have energy to make glucose, but what do plants use as their carbon source? CO_2, of course. You've probably heard of the term **carbon fixation**. All this means is that CO_2 from the air is converted into carbohydrates. This step occurs in the stroma of the leaf. The dark reactions are also called the **Calvin-Benson Cycle** (or just Calvin cycle).

Let's summarize the important facts about the dark reaction:

* The Calvin cycle occurs in the stroma of chloroplasts.
* ATP and NADPH from the light reaction are necessary for carbon fixation.
* CO_2 is fixed to form glucose.

The following table summarizes the stages of photosynthesis:

Stage of Photosynthesis	Location	Input	Output
Light-Dependent Reactions	Thylakoid membrane of chloroplasts Photosystem II (P680) Photosystem I (P700)	Photons H_2O	NADPH ATP O_2
Light-Independent Reactions Calvin Cycle	Stroma	3 CO_2 9 ATP 6 NADPH	sugar

CELLULAR RESPIRATION: THE SHORTHAND VERSION

In the shorthand version, cellular respiration looks something like this.

$$C_6H_{12}O_6 + 6O_2 \rightarrow 6CO_2 + 6H_2O + ATP$$

Notice that we've taken a sugar, perhaps a molecule of glucose, and combined it with oxygen to produce carbon dioxide, water, and energy in the form of our old friend, ATP. However, as you probably already know, the actual picture of what really happens is far more complicated.

Generally speaking, we can break cellular respiration down into two different approaches: aerobic respiration and anaerobic respiration. If ATP is made in the presence of oxygen, we call it **aerobic respiration**. If oxygen isn't present, we call it **anaerobic respiration**. Let's jump right in with aerobic respiration.

AEROBIC RESPIRATION

Aerobic respiration consists of four stages:

1. glycolysis
2. formation of acetyl-CoA
3. the Krebs (or citric acid) cycle
4. oxidative phosphorylation (or the electron transport chain + chemiosmosis)

In the first three stages, glucose is broken down and energy molecules are made. Some of these are ATP, and others are special electron carriers called **NADH** and **FADH$_2$**. In the fourth stage, the electron carriers unload their electrons, and the energy is eventually used to make even more ATP.

Stage 1: Glycolysis

The first stage begins with **glycolysis**, the splitting (*-lysis*) of glucose (*glyco-*). Glucose is a six-carbon molecule that is broken into two three-carbon molecules called **pyruvic acid**. This breakdown of glucose also results in the net production of two molecules of ATP.

$$\text{Glucose} + 2\ \text{ATP} + 2\text{NAD}^+ \rightarrow 2\ \text{Pyruvic acid} + 4\ \text{ATP} + 2\text{NADH}$$

Although we've written glycolysis as if it were a single reaction, this process doesn't occur in one step. In fact, it requires a sequence of enzyme-catalyzed reactions.

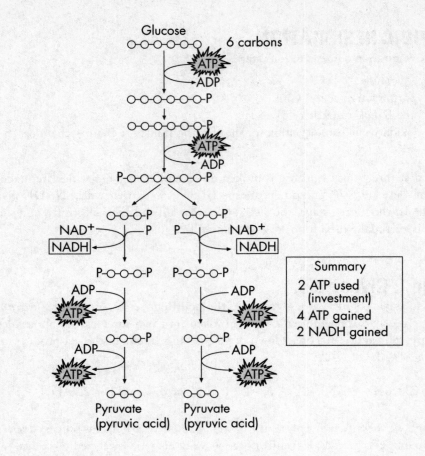

Glucose 6 carbons

Summary
2 ATP used (investment)
4 ATP gained
2 NADH gained

Pyruvate (pyruvic acid)   Pyruvate (pyruvic acid)

If you take a good look at the reaction above, you'll see two ATPs are needed to produce four ATP. You've probably heard the expression, "You have to spend money to make money." In biology, you have to invest ATP to make ATP: our investment of two ATP yielded four ATP, for a net gain of two. The ATP molecules generated in glycolysis are created from combining ADP and an inorganic phosphate with the help of an enzyme.

Glycolysis also creates two NADH, which result from the transfer of electrons to the carrier NAD⁺, which then becomes NADH. NAD⁺ and NADH are constantly being turned into each other as electrons are being carried and then unloaded.

There are four important tidbits to remember regarding glycolysis:

- occurs in the cytoplasm
- net of 2 ATPs produced
- 2 pyruvic acids formed
- 2 NADH produced

Stage 2: Formation of Acetyl-CoA

Pyruvic acid is transported to the mitochondrion. Each pyruvic acid (a three-carbon molecule) is converted to **acetyl-coenzyme A** (a two-carbon molecule, usually just called acetyl-CoA) and CO_2 is released.

2 Pyruvic acid + 2 Coenzyme A + 2NAD⁺ → 2 Acetyl-CoA + 2CO₂ + 2NADH

Are you keeping track of our carbons? We've now gone from two three-carbon molecules to two two-carbon molecules. The extra carbons leave the cell in the form of CO_2. Once again, two molecules of NADH are also produced. This process of turning pyruvic acid into acetyl-CoA is called the **pyruvate dehydrogenase complex**.

In the mitochondria, pyruvate is turned into acetyl-CoA and 1 NADH is made (double this if you are counting per glucose).

Stage 3: The Krebs Cycle

The next stage is the **Krebs cycle**, also known as the **citric acid cycle**. Each of the two acetyl coenzyme A molecules will enter the Krebs cycle, one at a time, and all the carbons will ultimately be converted to CO_2. This stage occurs in the **matrix** of the mitochondria.

The Krebs cycle begins with each molecule of acetyl-CoA produced from the second stage of aerobic respiration combining with **oxaloacetate**, a four-carbon molecule, to form a six-carbon molecule, **citric acid** or citrate (hence its name, the citric acid cycle).

Citrate gets turned into several other things, and because the cycle begins with a four-carbon molecule, oxaloacetate, it eventually gets turned back into oxaloacetate to maintain the cycle by joining with the next acetyl-CoA coming down the pipeline.

The Krebs cycle occurs in the mitochondrial matrix. It begins with acetyl-CoA joining with oxaloacetate to make citric acid and ends with oxaloacetate, 1 ATP, 3 NADH, and 1 FADH₂ (double this if you are counting per glucose).

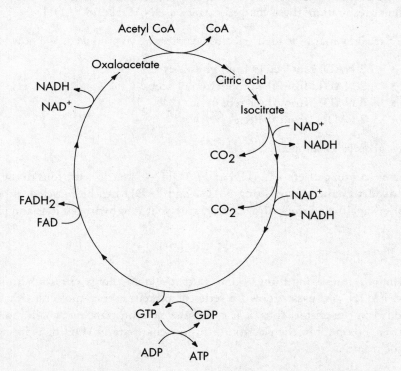

With each turn of the cycle, three types of energy are produced:

- 1 ATP
- 3 NADH
- 1 FADH$_2$

To figure out the total number of products per molecule of glucose, we simply double the number of products—after all, we started off the Krebs cycle with two molecules of acetyl-CoA for each molecule of glucose!

Now we're ready to tally up the number of ATP produced.

So far, we've made only four ATP—two ATP from glycolysis and two ATP from the Krebs cycle.

Although that seems like a lot of work for only four ATP, we have also produced hydrogen carriers in the form of 10 NADH and 2 FADH$_2$. These molecules will, in turn, produce lots of ATP.

Stage 4: Oxidative Phosphorylation

Electron Transport Chain

As electrons (and the hydrogen atoms to which they belong) are removed from a molecule of glucose, they carry with them much of the energy that was originally stored in their chemical bonds. These electrons—and their accompanying energy—are then transferred to readied hydrogen carrier molecules. In the case of cellular respiration, these charged carriers are NADH and FADH$_2$.

Let's see how many "loaded" electron carriers we've produced. We now have:

- 2 NADH molecules from glycolysis
- 2 NADH from the production of acetyl-CoA
- 6 NADH from the Krebs cycle
- 2 FADH$_2$ from the Krebs cycle

That gives us 12 altogether.

These electron carriers—NADH and FADH$_2$—"shuttle" electrons to the **electron transport chain**, the resulting NAD+ and FADH can be recycled to be used as carriers again, and the hydrogen atoms are split into hydrogen ions and electrons.

$$H_2 \rightarrow 2H^+ + 2e^-$$

Then, two interesting things occur. First, the high-energy electrons from NADH and FADH$_2$ are passed down a series of protein carrier molecules that are embedded in the cristae; thus, it is called the electron transport chain. Some of the carrier molecules in the electron transport chain are NADH dehydrogenase and cytochrome C.

Each carrier molecule hands down the electrons to the next molecule in the chain. The electrons travel down the electron transport chain until they reach the final electron acceptor, oxygen. Oxygen combines with these electrons (and some hydrogens) to form water. This explains the "aerobic" in aerobic respiration. If oxygen weren't available to accept the electrons, they wouldn't move down the chain at all, thereby shutting down the whole process of electron transport.

Oxygen is required as the final stop in the electron transport chain.

CHEMIOSMOSIS

At the same time that electrons are being passed down the electron transport chain, another mechanism is at work. Remember those hydrogen ions (also called *protons*) that split off from the original hydrogen atom? The energy released from the electron transport chain is used to pump hydrogen ions across the inner mitochondrial membrane from the matrix into the intermembrane space. The pumping of hydrogen ions into the intermembrane space creates a **pH gradient**, or **proton gradient**. The hydrogen ions really want to diffuse back into the matrix. The potential energy established in this gradient is responsible for the production of ATP. This pumping of ions and diffusion of ions to create ATP is **chemiosmosis**.

Brown fat, which is found in newborns and hibernating mammals, can also use the hydrogen gradient generated from the electron transport chain to generate heat.

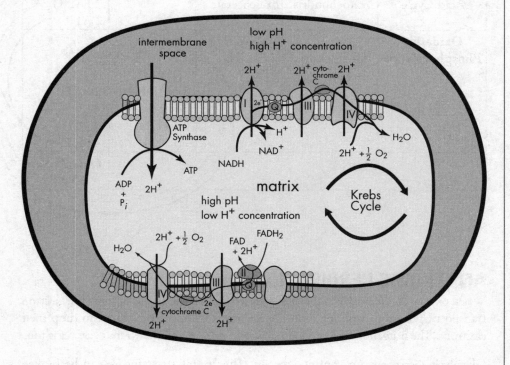

These hydrogen ions can diffuse across the inner membrane only by passing through channels called **ATP synthase**. Meanwhile, ADP and P_i are on the other side of these protein channels. The flow of protons through these channels produces ATP by combining ADP and P_i on the matrix side of the channel. Overall, this process is called **oxidative phosphorylation** because when electrons are given up it is called "oxidation" and then ADP is "phosphorylated" to make ATP.

You're also expected to know the following two things for the AP Biology Exam:

- Every NADH from glycolysis yields 1.5 ATP and every other NADH yields 2.5 ATP.
- Every $FADH_2$ yields 1.5 ATP.

For the AP Biology Exam, all you need to remember are the big steps and the overall outcome.

STAGES OF AEROBIC RESPIRATION				
Process	Location	Main Input	Main Output	Energy Output (per glucose)
Glycolysis	Cytoplasm	1 Glucose	2 pyruvates (2 net)	2 ATP 2 NADH
Formation of Acetyl-CoA	As pyruvate is transported into the mitochondria	2 pyruvates 2 Coenzyme A	2 Acetyl-CoA	2 NADH
Krebs or Citric Acid Cycle	Matrix of mitochondria	2 Acetyl-CoA oxaloacetate	oxaloacetate	6 NADH 2 $FADH_2$ 2 ATP
Oxidative Phosphorylation	Inner mitochondrial membrane	10 NADH 2 $FADH_2$ 2 O_2	Oxygen 2 NADH (× 1.5) 8 NADH (× 2.5) 2 $FADH_2$ (× 1.5)	3 ATP 20 ATP 3 ATP 4 ATP
				Overall Net: 30 ATP

ANAEROBIC RESPIRATION

When oxygen is not available, the anaerobic version of respiration occurs. The electron transport chain stops working, and the electron carriers have nowhere to drop their electrons. The mitochondrial production of acetyl-CoA and the Krebs cycle cease too.

Glycolysis, however, can continue to run. This means that glucose can be broken down to give net two ATP. Only two instead of 30!

Glycolysis also gives two pyruvates and two NADH. The pyruvate and the NADH make a deal with each other, and the pyruvate helps the NADH get recycled back into NAD+ and takes its electrons. The pyruvate turns into either **lactic acid** (in muscles) or **ethanol** (in yeast). Since these two things are toxic, this process, called **fermentation**, is only done in emergencies. Aerobic respiration is a better option.

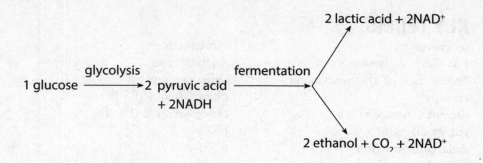

What types of organisms undergo fermentation?

- Yeast cells and some bacteria make ethanol and carbon dioxide.
- Other bacteria produce lactic acid.

Your Muscle Cells Can Ferment

Although human beings are aerobic organisms, they can actually carry out fermentation in their muscle cells. Have you ever had a leg cramp? If so, that cramp was possibly the consequence of anaerobic respiration.

When you exercise, your muscles require a lot of energy. To get this energy, they convert enormous amounts of glucose to ATP. But as you continue to exercise, your body doesn't get enough oxygen to keep up with the demand in your muscles. This creates an oxygen debt. What do your muscle cells do? They switch over to anaerobic respiration. Pyruvic acid produced from glycolysis is converted to lactic acid. As a consequence, the lactic acid causes your muscles to ache.

KEY TERMS

bioenergetics
First Law of Thermodynamics
Second Law of Thermodynamics
entropy
exergonic reaction
energy diagram
endergonic reaction
Gibbs free energy
enthalpy
spontaneous reactions
transition state
activation energy
enzymes
enzyme specificity
substrates
active site
enzyme-substrate complex
induced fit
cofactors
coenzymes
denatured
Q_{10}
allosteric sites
competitive inhibition
allosteric inhibitor
noncompetitive inhibition
cellular respiration
photosynthesis
light reactions
dark reactions
photons
stroma
grana
thylakoids
chlorophyll *a* and *b*

carotenoids
reaction center
antenna pigments
photosystem I (PS I)
photosystem II (PS II)
p680
p700
photophosphorylation
absorption spectrum
emission spectrum
photolysis
NADPH
carbon fixation
Calvin-Benson Cycle
aerobic respiration
anaerobic respiration
NADH
$FADH_2$
glycolysis
pyruvic acid
acetyl-coenzyme A (acetyl-CoA)
pyruvate dehydrogenase complex
Krebs cycle (or citric acid cycle)
matrix
oxaloacetate
citric acid
electron transport chain
chemiosmosis
cytochrome C
pH gradient
   (or proton gradient)
ATP synthase
oxidative phosphorylation
lactic acid
ethanol (or ethyl alcohol)
fermentation

Summary

○ Energy cannot be created or destroyed. There are endergonic reactions and exergonic reactions. The change in Gibbs free energy indicates whether the reaction will require energy or release energy.

○ The study of how cells accomplish biological processes is called bioenergetics.
- Chemical reactions are catalyzed by enzymes.
- Enzymes lower the activation energy of chemical reactions by stabilizing the transition state.
- Enzymes do not change the energy difference between reactants and products, but by lowering the activation energy, they facilitate chemical reactions to occur.

○ Enzymes are proteins that are highly specific for their substrate (reactant, which binds at the active site).

○ Enzymes have an optimal, narrow range of temperature and pH in which they have the highest rate of reaction. Outside this range, they may undergo denaturation and, therefore, no longer be active.

○ Enzyme activity may also be regulated or altered by allosteric/noncompetitive inhibitors, competitive inhibitors, and activators.

○ Enzymes may require coenzymes or cofactors to help catalyze reactions.

○ ATP is the universal energy molecule in cells. It is created via cellular respiration.

○ Photosynthesis is the process by which plants use energy from sunlight to make sugar.
- In the light-dependent reactions, the sunlight is absorbed by chlorophyll and the electrons are passed down a chain and the energy molecules ATP and NADPH are produced. Water is also split into hydrogen and oxygen.
- In the light-independent reactions, sugar is made in the Calvin-Benson Cycle using the energy from ATP and NADPH and the carbon from CO_2.

○ Cellular respiration consumes oxygen and produces a lot of ATP energy for the cell, using NADH and $FADH_2$ as electron carriers. There are four stages to cellular respiration:
- glycolysis
- formation of acetyl-CoA
- Krebs cycle
- oxidative phosphorylation

○ Anaerobic respiration occurs when cells lack oxygen to act as a final electron acceptor. In this process, the pyruvates generated by glycolysis are broken down by fermentation to produce lactic acid or ethanol and NAD^+, which permits glycolysis to continue and provide the 2 ATP it makes.

Chapter 6 Drill

Answers and explanations can be found in Chapter 16.

1. The mitochondrion is a critical organelle structure involved in cellular respiration. Below is a simple schematic of the structure of a mitochondrion. Which of the structural components labeled below in the mitochondrion is the primary location of the Krebs cycle?

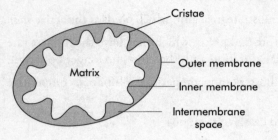

(A) Inner membrane
(B) Matrix
(C) Intermembrane space
(D) Outer membrane

2. Binding of Inhibitor Y as shown below inhibits a key catalytic enzyme by inducing a structural conformation change. Which of the following accurately describes the role of Inhibitor Y?

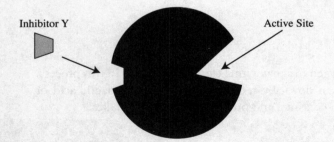

(A) Inhibitor Y competes with substrates for binding in the active site and functions as a competitive inhibitor.
(B) Inhibitor Y binds allosterically and functions as a competitive inhibitor.
(C) Inhibitor Y competes with substrates for binding in the active site and functions as a non-competitive inhibitor.
(D) Inhibitor Y binds allosterically and functions as a non-competitive inhibitor.

3. A single-step chemical reaction is catalyzed by the addition of an enzyme. Which of the reaction coordinate diagrams accurately shows the effect of the added enzyme (represented by the dashed line) to the reaction?

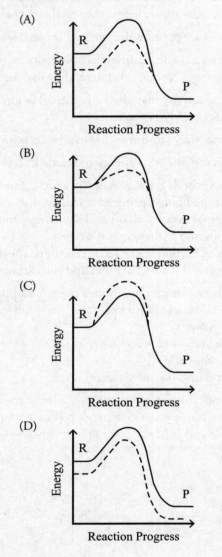

Questions 4-6 refer to the following diagram and paragraph.

Glycolysis (shown below) is a critical metabolic pathway that is utilized by nearly all forms of life. The process of glycolysis occurs in the cytoplasm of the cell and converts 1 molecule of glucose into 2 molecules of pyruvic acid.

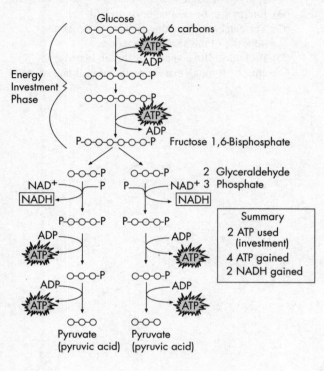

4. How many net ATP would be generated directly from glycolysis from the breakdown of 2 glucose molecules?

(A) 2
(B) 4
(C) 8
(D) 12

5. Glycolysis does not require oxygen to occur in cells. However, under anaerobic conditions, glycolysis normally requires fermentation pathways to occur to continue to produce ATP. Which best describes why glycolysis is dependent on fermentation under anaerobic conditions?

(A) Glycolysis requires fermentation to produce more glucose as a substrate.
(B) Glycolysis requires fermentation to synthesize lactic acid and restore NADH to NAD^+.
(C) Glycolysis requires fermentation to generate ATP molecules to complete the early steps of the pathway.
(D) Glycolysis requires fermentation to generate pyruvate for a later step in the pathway.

6. Which of the following most accurately describes the net reaction of glycolysis?

(A) It is an endergonic process because it results in a net increase in energy.
(B) It is an exergonic process because it results in a net increase in energy.
(C) It is an endergonic process because it results in a net decrease in energy.
(D) It is an exergonic process because it results in a net decrease in energy.

7. *Taq* polymerase, a DNA polymerase derived from thermophilic bacteria, is used in polymerase chain reactions (PCR) in the laboratory. During PCR, *Taq* catalyzes DNA polymerization, similar to how it would in bacteria. A normal PCR cycle is as follows:

1. Melting/Denaturing	95°C
2. Primer Annealing	50°C
3. Elongation of DNA	72°C
(repeat 20–30 cycles)	

Which of the following conditions likely describes the living environment of *Taq* bacteria?

(A) Freshwater with acidic pH
(B) Hydrothermal vents reaching temperatures between 70–75°C
(C) Hot springs of 40°C
(D) Tide pools with high salinity

8. The biggest difference between an enzyme-catalyzed reaction and an uncatalyzed reaction is

(A) the free energy between the reactants and the products
(B) the free energy difference between the reactants and the products does not change
(C) the catalyzed reaction would not occur without the enzyme
(D) the energy required to reach the transition state of the reaction

9. Two groups of cells were grown under identical conditions. Mitochondria from each group were isolated and half of them were placed in a low pH (approximately pH 6.8) and the other half were placed in a neutral pH. Small molecules were allowed to diffuse across the outer membrane via facilitated diffusion. Both samples were exposed to oxygen bubbles through the growth media. What would you expect to see in terms of ATP production in the sample of cells placed in a low pH, with respect to the control population?

(A) ATP production decreases.
(B) ATP production increases.
(C) ATP production stays the same.
(D) ATP production ceases entirely.

10. The Second Law of Thermodynamics states that entropy, or disorder, is constantly increasing in the universe for spontaneous processes. Therefore, how is it possible that organisms exist in very ordered states?

(A) The Second Law of Thermodynamics does not apply to biological life.
(B) Energy can be created or destroyed.
(C) The catabolic reactions are always equal and opposite of the anabolic reactions.
(D) Biological life actually creates an increase in entropy through dissemination of heat and waste.

REFLECT

Respond to the following questions:

- Which topics from this chapter do you feel you have mastered?

- Which content topics from this chapter do you feel you need to study more before you can answer multiple-choice questions correctly?

- Which content topics from this chapter do you feel you need to study more before you can effectively compose a free response?

- Was there any content that you need to ask your teacher or another person about?

Chapter 7
Molecular Biology

DNA: THE BLUEPRINT OF LIFE

All living things possess an astonishing degree of organization. From the simplest single-celled organism to the largest mammal, millions of reactions and events must be coordinated precisely for life to exist. This coordination is directed by DNA, which is the hereditary blueprint of the cell.

THE MOLECULAR STRUCTURE OF DNA

DNA is made up of repeated subunits of **nucleotides**. Each nucleotide has a **five-carbon sugar,** a **phosphate**, and a **nitrogenous base**. Take a look at the nucleotide below.

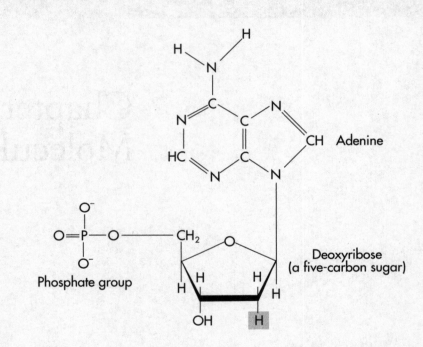

The name of the pentagon-shaped sugar in DNA is **deoxyribose**. Hence, the name *deoxyribo*nucleic acid. Notice that the sugar is linked to two things: a phosphate and a nitrogenous base. A nucleotide can have one of four different nitrogenous bases:

- **adenine**—a purine (double-ringed nitrogenous base)
- **guanine**—a purine (double-ringed nitrogenous base)
- **cytosine**—a pyrimidine (single-ringed nitrogenous base)
- **thymine**—a pyrimidine (single-ringed nitrogenous base)

Any of these four bases can attach to the sugar. As we'll soon see, this is extremely important when it comes to the message of the genetic code in DNA.

The nucleotides can link up in a long chain to form a single strand of DNA. Here's a small section of a DNA strand.

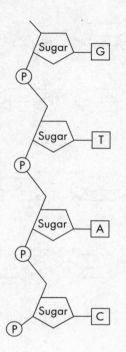

The nucleotides themselves are linked together by **phosphodiester bonds** between the sugars and the phosphates. This is called the sugar-phosphate backbone of DNA that serves as a scaffold for the bases.

Two DNA Strands

Each DNA molecule consists of two strands that wrap around each other to form a long, twisted ladder called a **double helix**. The structure of DNA was brilliantly deduced in 1953 by three scientists: **Watson, Crick,** and **Franklin.**

Now let's look at the way two DNA strands get together. Again, think of DNA as a ladder. The sides of the ladder consist of alternating sugar and phosphate groups, while the rungs of the ladder consist of pairs of nitrogenous bases.

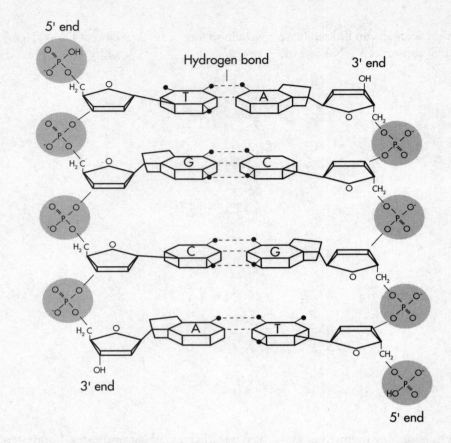

The nitrogenous bases pair up in a particular way. Adenine in one strand always binds to thymine (A–T or T–A) in the other strand. Similarly, guanine always binds to cytosine (G–C or C–G). This predictable matching of the bases is known as **base pairing**.

The two strands are said to be **complementary**. This means that if you know the sequence of bases in one strand, you'll know the sequence of bases in the other strand. For example, if the base sequence in one DNA strand is A–T–C, the base sequence in the complementary strand will be T–A–G.

The two DNA strands run in opposite directions. You'll notice from the figure above that each DNA strand has a 5' end and a 3' end, so-called for the carbon that ends the strand (i.e., the fifth carbon in the sugar ring is at the 5' end, while the third carbon is at the 3' end). The 5' end has a phosphate group, and the 3' end has an OH, or "hydroxyl," group. The 5' end of one strand is always opposite to the 3' end of the other strand. The strands are therefore said to be **antiparallel**.

The DNA strands are linked by **hydrogen bonds**. Two hydrogen bonds hold adenine and thymine together, and three hydrogen bonds hold cytosine and guanine together.

Before we go any further, let's review the base pairing in DNA:

- Adenine pairs up with thymine (A–T or T–A) by forming two hydrogen bonds.
- Cytosine pairs up with guanine (G–C or C–G) by forming three hydrogen bonds.

GENOME STRUCTURE

The order of the four base pairs in a DNA strand is the genetic code. Like a special alphabet in our cells, these four nucleotides spell out thousands of recipes. Each recipe is called a **gene**. The human genome has around 20,000 genes.

The recipes of the genes are spread out among the millions of nucleotides of DNA, and all of the DNA for a species is called its **genome**. Each separate chunk of DNA in a genome is called a **chromosome**.

Prokaryotes have one circular chromosome, and eukaryotes have linear chromosomes. In eukaryotes, the DNA is further structured, likely since linear chromosomes are more likely to get tangled. To keep it organized, the DNA is wrapped around proteins called histones, and then the histones are bunched together in groups called a **nucleosome**. Not all DNA is wound up equally, because it must be unwound in order to be read. How tightly DNA is packaged depends on the section of DNA and also what is going on in the cell at that time. Humans have 23 different chromosomes and they have two copies of each, so each human cell has 46 linear chromosome segments in the nucleus. The DNA of a cell is contained in structures called chromosomes. Chromosomes consist of DNA wrapped around proteins called **histones**. When the genetic material is in a loose form in the nucleus, it is called **euchromatin,** and its genes are active, or available for transcription. When the genetic material is fully condensed into coils, it is called **heterochromatin,** and its genes are generally inactive. Situated in the nucleus, chromosomes contain the recipe for all the processes necessary for life, including passing themselves and their information on to future generations. In this chapter, we'll look at precisely how they accomplish this.

DNA REPLICATION

We said in the beginning of this chapter that DNA is the hereditary blueprint of the cell. By directing the manufacture of proteins, DNA serves as the cell's blueprint. But how is DNA inherited? For the information in DNA to be passed on, it must first be copied. This copying of DNA is known as **DNA replication**.

Because the DNA molecule is twisted over on itself, the first step in replication is to unwind the double helix by breaking the hydrogen bonds. This is accomplished by an enzyme called **helicase**. The exposed DNA strands now form a *y*-shaped **replication fork.**

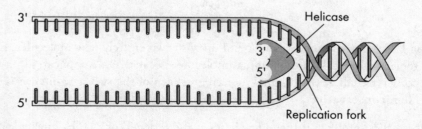

Now each strand can serve as a template for the synthesis of another strand. DNA replication begins at specific sites called **origins of replication**. Because the DNA helix twists and rotates during DNA replication, another class of enzymes, called DNA **topoisomerases**, cuts and rejoins the helix to prevent tangling. The enzyme that performs the actual addition of nucleotides to the freshly built strand is **DNA polymerase**. But DNA polymerase, oddly enough, can add nucleotides only to the 3' end of an existing strand. Therefore, to start off replication, an enzyme called **RNA primase** adds a short strand of RNA nucleotides called an **RNA primer**. After replication, the primer is degraded by enzymes and replaced with DNA so that the final strand contains only DNA filled with DNA.

One strand is called the **leading strand**, and it is made continuously. That is, the nucleotides are steadily added one after the other by DNA polymerase. The other strand—the **lagging strand**—is made discontinuously. Unlike the leading strand, the lagging strand is made in pieces of nucleotides known as **Okazaki fragments**. Why is the lagging strand made in small pieces?

Nucleotides are added only in the 5' to 3' direction since nucleotides can be added only to the 3' end of the growing chain. However, when the double-helix is "unzipped," one of the two strands is oriented in the opposite direction—3' to 5'. Because DNA polymerase doesn't work in this direction, the strand needs to be built in pieces. You can see in the figure below that the leading strand is being created toward the opening of the helix and the helix continually opens ahead of it to accommodate it. The lagging strand is built in the opposite direction of the way the helix is opening, so it can build only until it hits a previously built stretch. Once the helix unwinds a bit more, it can build another Okazaki fragment, and so on. These fragments are eventually linked together by the enzyme **DNA ligase** to produce a continuous strand. Finally, hydrogen bonds form between the new base pairs, leaving two identical copies of the original DNA molecule.

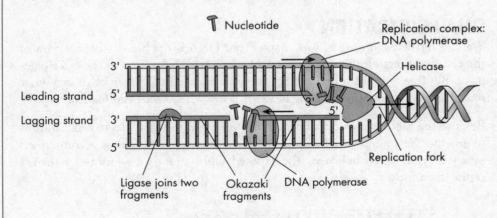

When DNA is replicated, we don't end up with two entirely new molecules. Each new molecule has half of the original molecule. Because DNA replicates in a way that conserves half of the original molecule in each of the two new ones, it is said to be **semiconservative**.

An interesting problem is that a few bases at the very end of a DNA molecule cannot be replicated because the DNA polymerase needs to bind. This means that every time replication occurs, the chromosome loses a few base pairs. The genome has compensated for this over time, by putting bits of unimportant (or at least less important) DNA at the ends of a molecule. These ends are called **telomeres**. They get shorter and shorter over time.

Many enzymes and proteins are involved in DNA replication. The ones you'll need to know for the AP Biology Exam are DNA helicase, DNA polymerase, DNA ligase, topoisomerase, and RNA primase:

- **Helicase** unwinds our double helix into two strands.
- **Polymerase** adds nucleotides to an existing strand.
- **Ligase** brings together the Okazaki fragments.
- **Topoisomerase** cuts and rejoins the helix.
- **RNA primase** catalyzes the synthesis of RNA primers.

Key History

The reason that we know that the DNA is the inheritable material is because of several key experiments. The first turning point was achieved by **Avery, MacLeod, and McCarty.** They isolated various cellular components from a dead virulent strain of bacteria. Then they followed up on a previous experiment by Griffiths and added each of these cellular components to a stain of living non-virulent bacteria. Only the DNA component of the deadly bacteria was able to change the second bacteria into a deadly strain capable of reproducing. This means that the DNA must be responsible for passing traits and it is inheritable.

The next important experiment was done by **Hershey-Chase**. They used bacteriophages, which are a special type of virus. They labelled the protein parts of some viruses with radiolabelled sulfur and they labelled the DNA parts of other viruses with radiolabelled phosphorus. When the viruses infected bacteria, only the labelled DNA was inside, but they were still able to replicate and make progeny viruses. Thus they proved that only DNA is required to pass on information. The protein was not required.

Central Dogma

DNA's main role is directing the manufacture of molecules that actually do the work in the body. DNA is just the recipe book, not the chef or the meal. When a DNA recipe is used, scientists say that that DNA is expressed.

The first step of DNA expression is to turn it into RNA. The RNA is then sent out into the cell and often gets turned into a protein. Proteins are the biggest group of "worker-molecules," and most expressed DNA turns into proteins. These proteins, in turn, regulate everything that occurs in the cell.

The process of making an RNA from DNA is called **transcription,** and the process of making a protein from an RNA is called **translation**. Transcription takes place in the nucleus (where the DNA is kept), and translation takes place in the cytoplasm.

Obviously, in prokaryotes, both occur in the cytoplasm because prokaryotes don't contain a nucleus; this allows transcription and translation to occur at the same time.

The flow of genetic information, the **Central Dogma** of Biology, is therefore:

$$\text{DNA} \xrightarrow[\text{in the nucleus}]{\text{transcription}} \text{RNA} \xrightarrow[\text{in the cytoplasm}]{\text{translation}} \text{Proteins}$$

Word Connection
A dogma is a core belief or set of ideas. Since this flow of information from DNA to RNA to protein is so important for understanding the workings of living things, it has been dubbed the Central Dogma of Biology.

RNA

Before we discuss transcription of RNA, let's talk about its structure. Although RNA is also made up of nucleotides, it differs from DNA in three ways.

1. RNA is single-stranded, not double-stranded.

2. The five-carbon sugar in RNA is **ribose** instead of deoxyribose.

3. The RNA nitrogenous bases are adenine, guanine, cytosine, and a different base called **uracil**. Uracil replaces thymine as adenine's partner.

Here's a table to compare DNA and RNA. Keep these differences in mind—the testing board loves to test you on them.

DIFFERENCES BETWEEN DNA AND RNA		
	DNA (double-stranded)	**RNA (single-stranded)**
Sugar:	deoxyribose	ribose
Bases:	adenine	adenine
	guanine	guanine
	cytosine	cytosine
	thymine	uracil

There are three main types of RNA: messenger RNA (mRNA), ribosomal RNA (rRNA), and transfer RNA (tRNA). All three types of RNA are key players in the synthesis of proteins:

Think Critically
Can you think of any exceptions to the Central Dogma? Do all genes get turned into RNA and then into proteins?

Hint: rRNA, tRNA, and RNAi never get turned into protein; they stay as RNA.

- **Messenger RNA (mRNA)** is a temporary RNA version of a DNA recipe that gets sent to the ribosome.
- **Ribosomal RNA (rRNA)**, which is produced in the nucleolus, makes up part of the ribosomes. You'll recall from our discussion of the cell in Chapter 5 that the ribosomes are the sites of protein synthesis. We'll see how they function a little later on.
- **Transfer RNA (tRNA)** shuttles amino acids to the ribosomes. It is responsible for bringing the appropriate amino acids into place at the appropriate time. It does this by reading the message carried by the mRNA.

There is also another class of RNA, called interfering RNAs, or **RNAi.** These are small snippets of RNA that are naturally made in the body or intentionally created by humans. These interfering RNAs, called siRNA and miRNA, can bind to specific sequences of RNA and mark them for destruction. This will be discussed more later in this chapter. Now that we know about the different types of RNA, let's see how they direct the synthesis of proteins.

Transcription

Transcription involves making an RNA copy of a bit of DNA code. The initial steps in transcription are similar to the initial steps in DNA replication. The obvious difference is that, whereas in replication we end up with a complete copy of the cell's DNA, in transcription we end up with only a tiny specific section copied into an mRNA. This is because only the bit of DNA that needs to be expressed will be transcribed. If you wanted to make a cake, but your cookbook couldn't leave your vault (AKA the nucleus), you wouldn't copy the entire cookbook! You would copy only the recipe for the cake.

Since each recipe is a gene, transcription occurs as-needed on a gene-by-gene basis.

The exception to this is prokaryotes because they will transcribe a recipe that can be used to make several proteins. This is called a **polycistronic** transcript. Eukaryotes tend to have one gene that gets transcribed to one mRNA and translated into one protein. Our transcripts are **monocistronic.**

Transcription involves three phases: initiation, elongation, and termination. As in DNA replication, the first initiation step in transcription is to unwind and unzip the DNA strands using helicase. Transcription begins at special sequences of the DNA strand called **promoters**. You can think of a promoter as a docking site or a runway. We will talk about how promoters are involved in regulating transcription later in the chapter.

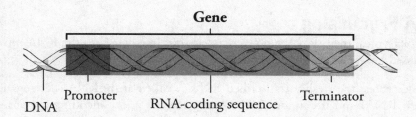

Gene

DNA | Promoter | RNA-coding sequence | Terminator

Because RNA is single-stranded, we have to copy only *one* of the two DNA strands. The strand that serves as the template is known as the **antisense strand,** the **non-coding strand,** or the **template strand**. The other strand that lies dormant is the **sense strand,** or the **coding strand.**

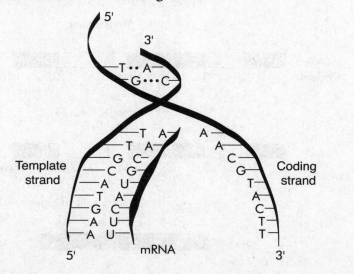

Just as DNA polymerase builds DNA, **RNA polymerase** builds RNA, and just like DNA polymerase, RNA polymerase adds nucleotides only to the 3' side (5' to 3'). This means that the RNA polymerase must bind to the 3' end of the template strand (which would be the 5' end of future mRNA).

RNA polymerase doesn't need a primer, so it can just start transcribing the DNA right off the bat. The promoter region is considered to be "upstream" of the actual coding part of the gene. This way the polymerase can get set up before the bases it needs to transcribe, like a staging area before an official parade starting point. The official starting point is called the **start site**.

When transcription begins, RNA polymerase travels along and builds an RNA that is complementary to the template strand of DNA. It is just like DNA replication except when DNA has an adenine, the RNA can't add a thymine since RNA doesn't have thymine. Instead the RNA adds a uracil.

Once the mRNA finishes adding on nucleotides and reaches the termination sequence, it separates from the DNA template, completing the process of transcription. The new RNA is now a transcript or a copy of the sequence of nucleotide based on the DNA strand. Note that the freshly synthesized RNA is complementary to the template strand, but it is identical to the coding strand (with the substitution of uracil for thymine).

RNA Processing

In prokaryotes, the mRNA is now complete, but in eukaryotes the RNA must be processed before it can leave the nucleus.

In eukaryotes, the freshly transcribed RNA is called an hnRNA (heterogeneous nuclear RNA) and it contains both coding regions and noncoding regions. The regions that express the code that will be turned into protein are **exons**. The noncoding regions in the mRNA are **introns**.

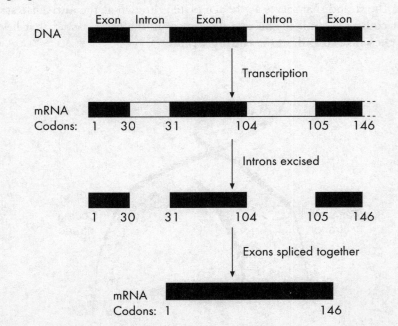

The introns—the intervening sequences—must be removed before the mRNA leaves the nucleus. This process, called **splicing**, is accomplished by an RNA-protein complex called a **spliceosome**. This process produces a final mRNA that is shorter than the transcribed RNA.

In addition to splicing, a **poly(A) tail** is added to the 3' end and a **5' GTP cap** is added to the 5' end.

Translation

Translation is the process of turning an mRNA into a protein. Remember, each protein is made of amino acids. The order of the mRNA nucleotides will be read in the ribosome in groups of three. Three nucleotides is called a **codon**. Each codon corresponds to a particular amino acid. The genetic code is redundant, meaning that certain amino acids are specified by more than one codon.

The mRNA attaches to the ribosome to initiate translation and "waits" for the appropriate amino acids to come to the ribosome. That's where tRNA comes in. A tRNA molecule has a unique three-dimensional structure that resembles a four-leaf clover:

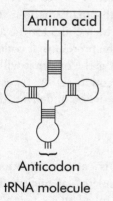

Amino acid

Anticodon

tRNA molecule

Bring on the Bio!
If you want extra practice, remember to check out our supplemental title, *550 AP Biology Practice Questions*.

One end of the tRNA carries an amino acid. The other end, called an **anticodon**, has three nitrogenous bases that can complementarily base pair with the codon in the mRNA. Usually, the normal rules of base pairing are set in stone, but tRNA anticodons can be a bit flexible when they bind with a codon on an mRNA, especially the 3 nucleotide in an anticodon. The third position is said to experience **wobble pairing**. Things that don't normally bind will pair up, like guanine and uracil.

Transfer RNAs are the "go-betweens" in protein synthesis. Each tRNA becomes charged and enzymatically attaches to an amino acid in the cell's cytoplasm and "shuttles" it to the ribosome. The charging enzymes involved in forming the bond between the amino acid and tRNA require ATP.

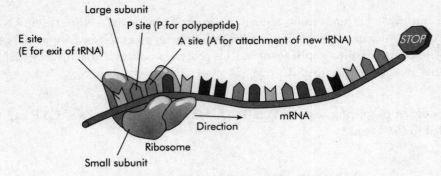

Note: Each site E, P, or A is a groove for the tRNA to fit in.

Translation also involves three phases: **initiation**, **elongation**, and **termination**. Initiation begins when a ribosome attaches to the mRNA.

What does the ribosome do? It helps the process along by holding everything in place while the tRNAs assist in assembling polypeptides.

Initiation

Ribosomes contain three binding sites: an **A site**, a **P site**, and an **E site**. The mRNA will shuffle through from A to P to E. As the mRNA codons are read, the polypeptide will be built. In all organisms, the **start codon** for the initiation of protein synthesis is A–U–G, which codes for the amino acid methionine. Proteins can have AUGs in the rest of the protein that code for methionine as well, but the special first AUG of an mRNA is the one that will kick off translation. The tRNA with the complementary anticodon, U–A–C, is methionine's personal shuttle; when the AUG is read on the mRNA, it delivers the methionine to the ribosome.

Elongation

The addition of amino acids is called elongation. Remember that the mRNA contains many codons, or "triplets," of nucleotide bases. As each amino acid is brought to the mRNA, it is linked to its neighboring amino acid by a peptide bond. When many amino acids link up, a polypeptide is formed.

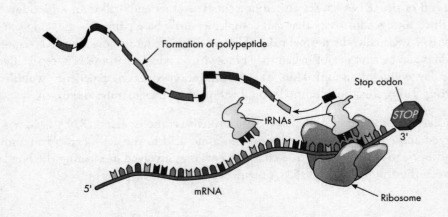

Termination

How does this process know when to stop? The synthesis of a polypeptide is ended by **stop codons**. A codon doesn't always code for an amino acid; there are three that serve as a stop codon. Termination occurs when the ribosome runs into one of these three stop codons.

How about a little review?

- In transcription, mRNA is created from a particular gene segment of DNA.
- In eukaryotes, the mRNA is "processed" by having its introns, or noncoding sequences, removed. A 5' cap and a 3' tail are also added.
- Now, ready to be translated, mRNA proceeds to the ribosome.
- Free-floating amino acids are picked up by tRNA and shuttled over to the ribosome, where mRNA awaits.
- In translation, the anticodon of a tRNA molecule carrying the appropriate amino acid base pairs with the codon on the mRNA.
- As new tRNA molecules match up to new codons, the ribosome holds them in place, allowing peptide bonds to form between the amino acids.
- The newly formed polypeptide grows until a stop codon is reached.
- The polypeptide or protein folds up and is released into the cell.

Gene Regulation

What controls gene transcription, and how does an organism express only certain genes? Regulation of gene expression can occur at different times. The largest point is before transcription, or **pre-transcriptional regulation**. It can also occur post-transcriptionally or post-translationally.

The start of transcription requires the DNA to be unwound and RNA polymerase to bind at the promoter. The process is usually a bit more complicated than that because there is a group of molecules called **transcription factors** that can either encourage or inhibit this from happening. This is often accomplished by making it easier or more difficult for the RNA polymerase to bind or to move to the start site.

The following examples are famous models of regulation. You do not need to memorize them, but you should be familiar with the big picture of how things can come together to regulate transcription. Most of what we know about gene regulation comes from our studies of *E. coli*. In bacteria, the region of bacterial DNA that regulates gene expression is called an **operon**. One of the best-understood operons is the *lac* operon, which controls expression of the enzymes that break down lactose.

The operon consists of four major parts: structural genes, the promoter genes, the operator, and the regulatory gene:

- **Structural genes** are genes that code for enzymes needed in a chemical reaction. These genes will be transcribed at the same time to produce particular enzymes. In the *lac* operon, three enzymes (beta galactosidase, galactose permease, and thiogalactoside transacetylase) involved in digesting lactose are coded for.

- The **promoter gene** is the region where the RNA polymerase binds to begin transcription.

- The **operator** is a region that controls whether transcription will occur; this is where the repressor binds.

- The **regulatory gene** codes for a specific regulatory protein called the repressor. The repressor is capable of attaching to the operator and blocking transcription. If the repressor binds to the operator, transcription will not occur. On the other hand, if the repressor does not bind to the operator, RNA polymerase moves right along the operator and transcription occurs. In the *lac* operon, the **inducer**, lactose, binds to the repressor, causing it to fall off the operator, and "turns on" transcription.

The following two diagrams show the lac operon when lactose is absent (A) and when lactose is present (B).

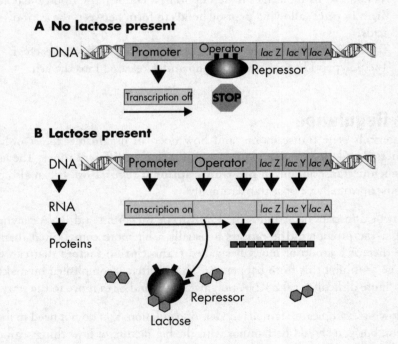

Other operons, such as the *trp* operon, operate in a similar manner except that this mechanism is continually "turned on" and is "turned off" only in the presence of high levels of the amino acid, tryptophan. Tryptophan is a product of the pathway that codes for the *trp* operon. When tryptophan combines with the *trp* repressor protein, it causes the repressor to bind to the operator, which turns the operon "off," thereby blocking transcription. In other words, a high level of tryptophan acts to repress the further synthesis of tryptophan.

The following two diagrams show the *trp* operon when tryptophan is absent (C) and when tryptophan is present (D).

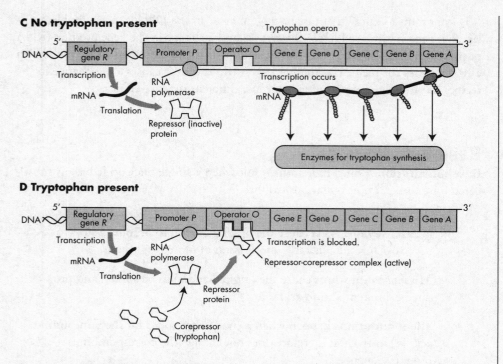

C No tryptophan present

D Tryptophan present

Post-transcriptional regulation occurs when the cell creates an RNA, but then decides that it should not be translated into a protein. This is where RNAi comes into play. RNAi molecules can bind to an RNA via complementary base pairing. This creates a double-stranded RNA (remember RNA is usually single stranded). When a double-stranded RNA is formed, this signals to some special destruction machinery that the RNA should be destroyed. This would prevent it from going on to be translated.

Post-translational regulation can also occur. This would be if a cell has already made a protein, but it doesn't need to use the protein yet. This is an especially common regulation for enzymes because when the cell needs them, it needs them ASAP. It is easier to make them ahead of time and then just turn them on or off as needed. This can involve binding with other proteins, phosphorylation, pH changes, cleavage, etc. Remember, even if a protein is made, it might need other things to be made (or not made) in order for the protein to be functional.

MUTATIONS

A **mutation** is a error in the genetic code. Mutations can occur because DNA is damaged and cannot be repaired or because DNA damage is repaired incorrectly. Damage can be caused by chemicals or radiation. It can also occur when a DNA polymerase or an RNA polymerase makes a mistake. DNA polymerases have proofreading abilities, but RNA polymerases do not. This is because RNA is only a temporary molecule and if a mistake is made, it is not as problematic. DNA is important now and in the future since it is passed on from cell to cell in somatic cells and from parent to offspring. If the mistake occurs in a germline cell, that will become a gamete. Mistakes in DNA can last forever.

It is important to know that just having an error in the DNA is not a problem unless that gene is expressed AND the error causes a change in the gene product (RNA or protein). Think of it like an error in a recipe in a cookbook; an error won't harm anyone unless the recipe is actually made AND the error causes a big change in the recipe. As we will see, a mutation can cause different types of effects.

Base Substitution

Base substitution (point) mutations result when a single nucleotide base is substituted for another. Point mutations can be:

- **Nonsense mutations** cause the original codon to become a stop codon, which results in early termination of protein synthesis.

- **Missense mutations** cause the original codon to be altered and produce a different amino acid.

- **Silent mutations** happen when a codon that codes for the same amino acid is created and therefore does not change the corresponding protein sequence.

There are two types of mutations: base substitutions and gene rearrangements.

Gene Rearrangements

Gene rearrangements involve DNA sequences that have deletions, duplications, inversions, and translocations.

1. **Insertions** and **deletions** result in the gain or loss, respectively, of DNA or a gene. They can involve either the addition (insertion) or removal (deletion) of as little as a single base or much larger sequences of DNA. Insertions and deletions may have devastating consequences during protein translation because the introduction or deletion of bases often results in a change in the sequence of codons used by the ribosome (called a **frameshift mutation**) to synthesize a polyprotein.

    Example:

Original DNA Sequence:	ATG	TAT	AAA	CAT	TGA
Original Polyprotein:	Met	Tyr	Lys	His	Stop
DNA Sequence after Insertion:	ATG	CTA	TAA	ACA	TTG A
Polyprotein Sequence after Insertion:	Met	Leu	Stop	-	-

    In this example, the insertion of an additional cytosine nucleotide resulted in a frameshift and a premature stop codon.

2. **Duplications** can result in an extra copy of genes and are usually caused by unequal crossing-over during meiosis or chromosome rearrangements. This may result in new traits because one copy can maintain the gene's original function and one copy may evolve a new function.

3. **Inversions** can result when changes occur in the orientation of chromosomal regions. This may cause harmful effects if the inversion involves a gene or an important regulatory sequence.

4. **Translocations** occur when two different chromosomes (or a single chromosome in two different places) break and rejoin in a way that causes the DNA sequence or gene to be lost, repeated, or interrupted.

BIOTECHNOLOGY

Recombinant DNA

Scientists have learned how to harness the Central Dogma in order to research and cure diseases. **Recombinant DNA** is generated by combining DNA from multiple sources to create a unique DNA molecule that is not found in nature. A common application of recombinant DNA technology is the introduction of a eukaryotic gene of interest (such as insulin) into a bacterium for production. In other words, bacteria can be hijacked and put to work as little protein factories. This branch of technology that produces new organisms or products by transferring genes between cells is called **genetic engineering**.

Polymerase Chain Reaction (PCR)

A few years ago, it would take weeks of tedious experiments to identify and study specific genes. Today, thanks to **polymerase chain reaction (PCR)**, we are able to make billions of identical copies of genes within a few hours. To do PCR, the process of DNA replication is slightly modified. In a small PCR tube, DNA, specifically designed primers, a powerful DNA polymerase, and lots of DNA nucleotides (A's, C's, G's, and T's) are mixed together.

In a PCR machine, or **thermocycler**, the tube is heated, cooled, and warmed many times. Each time the machine is heated, the hydrogen bonds break, separating the double-stranded DNA. As it is cooled, the primers bind to the sequence flanking the region of the DNA we want to copy. When it is warmed, the polymerase binds to the primers on each strand and adds nucleotides on each template strand. After this first cycle is finished, there are two identical double-stranded DNA molecules. When the second cycle is completed, these two double-stranded DNA segments will have been copied into four. The process repeats itself over and over, creating as much DNA as needed.

Today, a thermocycler is commonplace in science labs. It is regularly used to study small amounts of DNA from crime scenes, determine the origin of our foods, detect diseases in animals and humans, and better understand the inner workings of our cells.

Transformation

Insulin, the protein hormone that lowers blood sugar levels, can now be made for medical purposes by bacteria. Yes, bacteria can be induced to use the universal DNA code to transcribe and translate a human gene! The process of giving bacteria foreign DNA is called **transformation**.

Genes of interest are first placed into a small circular DNA molecule called a **plasmid**. Often, a plasmid is pre-designed to have some special helpful genes in it, and together, these are called a **vector**. Plasmid vectors often contain genes for antibiotic resistance and restriction sites. Plasmids and the gene of interest are cut with the same **restriction enzyme** (restriction enzymes cut DNA at very specific pre-determined sequences), creating compatible **sticky ends**. When placed together, the gene is inserted into the plasmid creating recombinant DNA.

The bacteria are then transformed (given the recombinant DNA to take up). In most AP Biology courses, this is done by the heat shock method.

Not all the bacteria will be transformed, but the ones that did not take up the plasmid are not needed. This is where the antibiotic resistance gene on the vector comes in. By growing all the bacteria in the presence of an antibiotic, only those with the resistance gene (AKA those that have been transformed) will survive.

This laboratory technique has not only been used to safely mass-produce important proteins used for medicine, such as insulin, it also plays an important role in the study of gene expression. A technique similar to transformation is **transfection**, which is putting a plasmid into a eukaryotic cell, rather than a bacteria cell. This is a bit trickier than transformation. Because eukaryotic cells do not grow like bacteria do, transfection is not as useful for making large quantities of something, but sometimes it is important to use eukaryotic cells.

Gel Electrophoresis

An important part of DNA technology is the ability to observe differences in different preparations of DNA.

The DNA fragments can be separated according to their molecular weight using **gel electrophoresis**. Because DNA and RNA are negatively charged, they migrate through the gel toward the positive pole of the electrical field. The smaller the fragments, the faster they move through the gel. Restriction enzymes are also used to create a molecular fingerprint. The patterns created after cutting with restriction enzymes are unique for each person because each person's DNA is slightly different. Some people might have sequences that are cut many times, resulting in many tiny fragments, and others might have sequences that are cut infrequently, producing a few large fragments. When restriction fragments between individuals of the same species are compared, the fragments differ in length because of polymorphisms, which are slight differences in DNA sequences. These fragments are called **restriction fragment length polymorphisms**, or RFLPs. In **DNA fingerprinting**, RFLPs produced from DNA left at a crime scene are compared to RFLPs from the DNA of suspects.

KEY TERMS

nucleotide
five-carbon sugar
phosphate
nitrogenous base
deoxyribose
adenine
guanine
cytosine
thymine
phosphodiester bonds
double helix
Watson, Crick, and Franklin
base pairing
complementary
antiparallel
hydrogen bonds
gene
genome
chromosome
nucleosome
histone
euchromatin
heterochromatin
DNA replication
helicase
replication fork
origins of replication
topoisomerase
DNA polymerase
RNA primase
RNA primer
leading strand
lagging strand
Okazaki fragments
DNA ligase
semiconservative replication
telomeres
Avery, MacLeod, and McCarty
Hershey-Chase
transcription
translation
Central Dogma

ribose
uracil
messenger RNA (mRNA)
ribosomal RNA (rRNA)
transfer RNA (tRNA)
RNA interference
   (RNAi)/silencing RNA (siRNA)
polycistronic
monocistronic
promoter
antisense/non-coding/template
   strand
sense/coding strand
RNA polymerase
start site
exons
introns
splicing
spliceosome
poly(A) tail
5' GTP cap
codon
anticodon
wobble pairing
initiation
elongation
termination
A site, P site, E site
start codon
stop codons
pre-transcriptional regulation
transcription factors
operon
structural genes
promoter genes
operator
inducer
regulatory gene
post-transcriptional regulation
post-translational regulation
mutation
base substitution

nonsense mutation
missense mutation
silent mutation
insertions
deletions
frameshift mutation
duplications
inversions
translocations
recombinant DNA
genetic engineering
polymerase chain reaction (PCR)
thermocycler
transformation
plasmid
vector
restriction enzyme
sticky end
transfection
gel electrophoresis
restriction fragment length
   polymorphism (RFLPs)
DNA fingerprinting

Summary

- o DNA is the genetic material of the cell.

- o DNA structure is composed of two antiparallel complementary strands, including:
  - Deoxyribose sugar and phosphate backbone connected by phosphodiester bonds with nitrogenous bases hydrogen bonded together pointing toward the middle. The two strands are oriented antiparallel with 3' and 5' ends and twist into a double helix.
  - Adenine always pairs with thymine, and guanine always pairs with cytosine.

- o All the DNA in a cell is the genome. It is divided up into chromosomes, which are divided up into genes that each encode a particular genetic recipe.

- o DNA replication provides two copies of the DNA for cell division, which involves the following steps:
  - Helicase begins replication by unwinding DNA and separating the strands at the origin of replication.
  - Topoisomerases reduce supercoiling ahead of the replication fork.
  - RNA primase places an RNA primer down on the template strands.
  - DNA polymerase reads the DNA base pairs of a strand as a template and lays down complementary nucleotides to generate a new complementary partner strand.
  - Replication proceeds continuously on the leading strand.
  - The lagging strand forms with Okazaki fragments, which must be joined later by DNA ligase.

- o Transcription involves building RNA from DNA.
  - RNA polymerase reads one of the DNA strands (the anti-sense/non-coding/template strand) and adds RNA nucleotides to generate a single strand of mRNA that will be identical to the other strand of DNA (the coding/sense strand) except it will have uracil instead of thymine.
  - mRNA, rRNA, tRNA, and RNAi are all types of RNA that can be created.
  - mRNA gets processed in eukaryotic cells: introns are removed (spliced out), and a 5' GTP cap and a 3' poly(A) tail are added.

- o Translation occurs in the cytoplasm as the mRNA carries the message to the ribosome.
  - The mRNA passes through three sites on the ribosome, the A-site, the P-site, and the E-site.
  - The mRNA is read in triplets of three nucleotides, called codons. Each tRNA has a region called the anticodon that is complementary to a codon. Sometimes the pairing is not the normal base-pairing. This is called wobble pairing.
  - When a tRNA binds, it brings the corresponding amino acid to add to the growing polypeptide.

○ Gene expression is regulated primarily by transcription factors influencing transcription (pre-transcriptional regulation), RNAi after transcription (post-transcriptional regulation), and by various things after translation (post-translational regulation). This regulation is dynamic and can either increase or decrease gene expression, RNA levels, and protein levels according to the needs of the cell.

○ Mutations can result from changes in the DNA message or the mRNA message.

○ Mutations can be small (single nucleotide swaps, additions, or deletions) or large (big chunks or even entire chromosomes that are swapped, duplicated, or deleted).

○ Some examples of biotechnology are:
  • recombinant DNA
  • polymerase chain reaction (PCR)
  • transformation of bacteria
  • gel electrophoresis

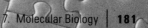

Chapter 7 Drill

Answers and explanations can be found in Chapter 16.

1. A geneticist has discovered a yeast cell, which encodes a DNA polymerase that may add nucleotides in both the 5′ to 3′ and 3′ to 5′ directions. Which of the following structures would this cell not likely generate during DNA replication?

    (A) RNA primers
    (B) Okazaki fragments
    (C) Replication fork
    (D) Nicked DNA by topoisomerases

2. A eukaryotic gene, which does not normally undergo splicing, was exposed to benzopyrene, a known carcinogen and mutagen. Following exposure, the protein encoded by the gene was shorter than before exposure. Which of the following types of genetic rearrangements or mutations was likely introduced by the mutagen?

    (A) Silent mutation
    (B) Missense mutation
    (C) Nonsense mutation
    (D) Duplication

3. DNA replication occurs through a complex series of steps involving several enzymes. Which of the following represents the correct order beginning with the earliest activity of enzymes involved in DNA replication?

    (A) Helicase, ligase, RNA primase, DNA polymerase
    (B) DNA polymerase, RNA primase, helicase, ligase
    (C) RNA primase, DNA polymerase, ligase, helicase
    (D) Helicase, RNA primase, DNA polymerase, ligase

4. If a messenger RNA codon is UAC, which of the following would be the complementary anticodon triplet in the transfer RNA?

    (A) ATG
    (B) AUC
    (C) AUG
    (D) ATT

5. During post-translational modification, the polypeptide from a eukaryotic cell typically undergoes substantial alteration that results in

    (A) excision of introns
    (B) addition of a poly(A) tail
    (C) formation of peptide bonds
    (D) a change in the overall conformation of a polypeptide

6. Which of the following represents the maximum number of amino acids that could be incorporated into a polypeptide encoded by 21 nucleotides of messenger RNA?

    (A) 3
    (B) 7
    (C) 21
    (D) 42

7. A researcher uses molecular biology techniques to insert a human lysosomal membrane protein into bacterial cells to produce large quantities of this protein for later study. However, only small quantities of this protein result in these cells. What is a possible explanation for this result?

    (A) The membrane protein requires processing in the ER and Golgi, which are missing in the bacterial cells.
    (B) Bacteria do not make membrane proteins.
    (C) Bacteria do not use different transcription factors than humans, so the gene was not expressed.
    (D) Bacteria do not have enough tRNAs to make this protein sequence.

8. *Bam*HI is a restriction enzyme derived from *Bacillus amyloliquefaciens* that recognizes short palindromic sequences in DNA. When the enzyme recognizes these sequences, it cleaves the DNA. What purpose would restriction enzymes have in a bacterium like *Bacillus*?

(A) They are enzymes that no longer have a purpose because evolution has produced better enzymes.
(B) They destroy extra DNA that results from errors in binary fission.
(C) They protect *Bacillus* from invading DNA due to viruses.
(D) They prevent, or restrict, DNA replication when the cell isn't ready to copy its DNA.

9. Viruses and bacteria have which of the following in common?

(A) Ribosomes
(B) Nucleic acids
(C) Flagella
(D) Metabolism

10. Griffith was a researcher who coined the term *transformation* when he noticed that incubating nonpathogenic bacteria with heat-killed pathogenic bacteria produced bacteria that ultimately became pathogenic, or deadly, in mice. What caused the transformation in his experiment?

(A) DNA from the nonpathogenic bacteria revitalized the heat-killed pathogenic bacteria.
(B) Protein from the pathogenic bacteria was taken up by the nonpathogenic bacteria.
(C) DNA from the pathogenic bacteria was taken up by the nonpathogenic bacteria.
(D) DNA in the nonpathogenic bacteria turned into pathogenic genes in the absence of pathogenic bacteria.

11. A biologist systematically removes each of the proteins involved in DNA replication to determine the effect each has on the process. In one experiment, after separating the strands of DNA, she sees many short DNA/RNA fragments as well as some long DNA pieces. Which of the following is most likely missing?

(A) Helicase
(B) DNA polymerase
(C) DNA ligase
(D) RNA primase

REFLECT

Respond to the following questions:

- Which topics from this chapter do you feel you have mastered?

- Which content topics from this chapter do you feel you need to study more before you can answer multiple-choice questions correctly?

- Which content topics from this chapter do you feel you need to study more before you can effectively compose a free response?

- Was there any content that you need to ask your teacher or another person about?

Chapter 8
Cell Reproduction

CELL DIVISION

Every second, thousands of cells are dying throughout our bodies. Fortunately, the body replaces them at an amazing rate. In fact, epidermal cells, or skin cells, die off and are replaced so quickly that the average 18-year-old grows an entirely new skin every few weeks. The body keeps up this unbelievable rate thanks to the mechanisms of **cell division**.

This chapter takes a closer look at how cells divide. But remember, cell division is only a small part of the life cycle of a cell. Most of the time, cells are busy carrying out their regular activities. There are also some types of cells that are nondividing. Often, these are highly specialized cells that are created from a population of less specialized cells. The body continually makes them as needed, but they do not directly replicate themselves. Red blood cells are an example of nondividing cells. Some other cells are just temporarily nondividing. They enter a phase called G_0, where they hang out until they get a signal to reenter the normal cell cycle.

THE CELL CYCLE

Special gametes split their time between interphase and meiosis.

Every cell has a life cycle—the period from the beginning of one division to the beginning of the next. The cell's life cycle is known as the **cell cycle**. The cell cycle is divided into two periods: **interphase** and **mitosis**. Take a look at the cell cycle of a typical cell.

Notice that most of the life of a cell is spent in interphase.

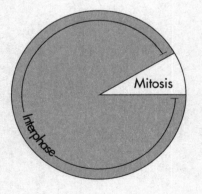

INTERPHASE: THE GROWING PHASE

Interphase is the time span from one cell division to another. We call this stage interphase (*inter-* means between) because the cell has not yet started to divide. Although biologists sometimes refer to interphase as the "resting stage," the cell is definitely not inactive. During this phase, the cell carries out its regular activities. All the proteins and enzymes it needs to grow are produced during interphase.

The Three Stages of Interphase

Interphase can be divided into three stages: **G₁**, **S**, **G₂**.

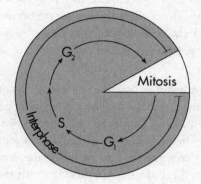

The most important phase is the S phase. That's when the cell replicates its genetic material. The first thing a cell has to do before undergoing mitosis is to duplicate all of its chromosomes, which contain the organism's DNA "blueprint." During interphase, every single chromosome in the nucleus is duplicated.

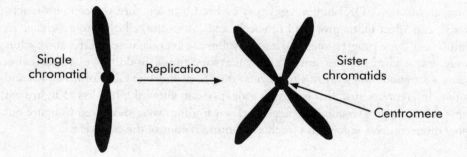

You'll notice that the original chromosome and its duplicate are still linked, like conjoined twins. These identical strands of DNA are now called **sister chromatids**. The chromatids are held together by a structure called the **centromere**. You can think of each chromatid as a chromosome, but because they remain attached, they are called chromatids instead. To be called a chromosome, they each need to have their own centromere. Once the chromatids separate, they will be full-fledged chromosomes.

Cell Cycle Regulation

We've already said that replication occurs during the S phase of interphase, so what happens during G₁ and G₂? During these stages, the cell produces proteins and enzymes. For example, during G₁, the cell produces all of the enzymes required for DNA replication (DNA helicase, DNA polymerase, and DNA ligase, as we learned in Chapter 7). By the way, *G* stands for "gap," but we can also associate it with "growth." These three phases are highly regulated by checkpoints and special proteins called **cyclins** and **cyclin-dependent kinases** (CDKs).

Cell cycle checkpoints are control mechanisms that make sure cell division is happening properly in eukaryotic cells. They monitor the cell to make sure it is ready to progress through the cell cycle and stop progression if it is not. In eukaryotes, checkpoint pathways function mainly at phase boundaries (such as the G_1/S transition and the G_2/M transition). Checkpoint pathways are an example of an important cell signaling pathway inside of cells.

For example, if the DNA genome has been damaged in some way, the cell should not divide. If it does, damaged DNA will be passed onto daughter cells, which can be disastrous. When damaged DNA is found, checkpoints are activated and cell cycle progression stops. The cell uses the extra time to repair damage in DNA. If the DNA damage is so extensive that it cannot be repaired, the cell can undergo **apoptosis**, or programmed cell death. Apoptosis cannot stop once it has begun, so it is a highly regulated process. It is an important part of normal cell turnover in multicellular organisms.

Cell cycle checkpoints control cell cycle progression by regulating two families of proteins: cyclin-dependent kinases (CDKs) and cyclins. To induce cell cycle progression, an inactive CDK binds a regulatory cyclin. Once together, the complex is activated, can affect many proteins in the cell, and causes the cell cycle to continue. To inhibit cell cycle progression, CDKs and cyclins are kept separate. CDKs and cyclins were first studied in yeast, unicellular eukaryotic fungi. Budding yeast and fission yeast were good model organisms because they have only one CDK protein each and a few different cyclins. This relatively simple system allowed biologists to figure out how cell cycle progression is controlled. Their findings were then used to figure out how more complex eukaryotes (such as mammals) control the cell cycle.

Cancer

A cell losing control of the cell cycle can have disastrous consequences. For example, a mutation in a protein that normally controls progression through the cell cycle can result in unregulated cell division and cancer. **Cancer** means "crab," as in the zodiac sign. The name comes from the observation that malignant tumors grow into surrounding tissue, embedding themselves like clawed crabs. Cancer occurs when normal cells start behaving and growing very abnormally and spread to other parts of the body.

Oncogenes and tumor suppressors can both cause cancer, but oncogenes do it by stepping on the cell cycle gas pedal, while tumor suppressors do it by removing the brake pedal.

Mutated genes that induce cancer are called **oncogenes** ("onco-" is a prefix denoting cancer). Normally, these genes are required for proper growth of the cell and regulation of the cell cycle. Oncogenes, then, are genes that can convert normal cells into cancerous cells. Often these are abnormal or mutated versions of normal genes. The normal healthy version is called a proto-oncogene.

Tumor suppressor genes produce proteins that prevent the conversion of normal cells into cancer cells. They can detect damage to the cell and work with CDK/cyclin complexes to stop cell growth until the damage can be repaired. They can also trigger apoptosis if the damage is too severe to be repaired.

Cell cycle checkpoints and apoptosis have been intimately linked with cancer. In order for a normal cell to become a cancer cell, it must override cell cycle checkpoints and grow in an unregulated way. It must also avoid cell death. Tumors often accumulate DNA damage. Cancer treatments target these changes, because they are what make a cancer cell different from a normal cell.

Let's recap:

- The cell cycle consists of two things: interphase and mitosis.
- During the S phase of interphase, the chromosomes replicate.
- Growth and preparation for mitosis occur during the G_1 and G_2 stages of interphase.
- Cell cycle checkpoints make sure the cell is ready to continue through the cell cycle.
- CDK and cyclin proteins work together to promote cell cycle progression.
- Oncogenes promote cell growth and tumor suppressor genes inhibit cell growth.

MITOSIS: THE DANCE OF THE CHROMOSOMES

Once the chromosomes have replicated, the cell is ready to begin mitosis. Mitosis is the period when the cell divides. Mitosis consists of a sequence of four stages: **prophase**, **metaphase**, **anaphase**, and **telophase**. It is not important to memorize the name of each phase, but it is important that you know the basic order of operations for mitosis.

Stage 1: Prophase: Prep (the cell prepares to divide)

One of the first signs of prophase is the disappearance of the nucleolus (ribosome making area) and the nuclear envelope. In prophase, the chromosomes thicken, forming coils upon coils, and become visible. The dense chromosomes are called **chromatin.** (During interphase, the chromosomes are not visible.)

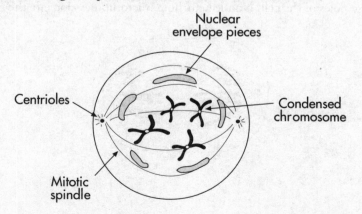

Now the cell has plenty of room to "sort out" the chromosomes. Remember centrioles? During prophase, these cylindrical bodies found within microtubule organizing centers (MTOCs) start to move away from each other, toward opposite ends of the cell. The centrioles will spin out a system of microtubules known as the **spindle fibers**. These spindle fibers will attach to a structure on each chromatid called a **kinetochore**. The kinetochores are part of the centromere.

Stage 2: Metaphase: Middle (chromosomes align in the middle)

The next stage is called metaphase. The chromosomes now begin to line up along the equatorial plane, or the **metaphase plate**, of the cell. That's because the spindle fibers are attached to the kinetochore of each chromatid.

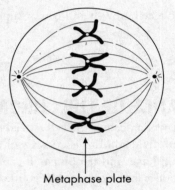

Metaphase plate

Stage 3: Anaphase: Apart (the chromatids are pulled apart)

During anaphase, the sister chromatids of each chromosome separate at the centromere and migrate to opposite poles. The chromatids are pulled apart by the microtubules, which begin to shorten. Each half of a pair of sister chromatids now moves to opposite poles of the cell. Non-kinetochore microtubules elongate the cell.

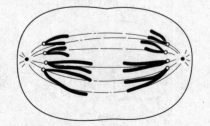

Stage 4: Telophase: Two (the cell completes splitting in two)

The final phase of mitosis is telophase. A nuclear membrane forms around each set of chromosomes and the nucleoli reappear.

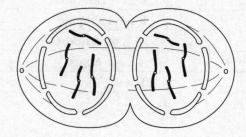

The nuclear membrane is ready to divide. Now it's time to split the cytoplasm in a process known as **cytokinesis**. Look at the figure below and you'll notice that the cell has begun to split along a **cleavage furrow** (which is produced by actin microfilaments).

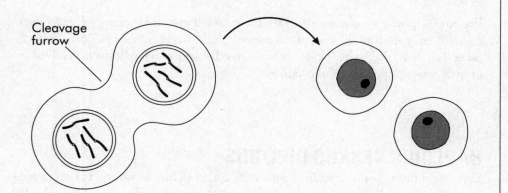

Cleavage furrow

A cell membrane forms about each cell, and the cells split into two distinct daughter cells. The division of the cytoplasm yields two daughter cells.

Here's one thing to remember: cytokinesis occurs differently in plant cells. The cell doesn't form a cleavage furrow. Instead, a partition called a **cell plate** forms down the middle region.

Stage 5: Interphase

Once the daughter cells are produced, they reenter the initial phase—interphase—and the whole process starts over. The cell goes back to its original state. Once again, the chromosomes become invisible, and the genetic material is called chromatin again.

But How Will I Remember All That?

For mitosis, you may already have your own mnemonic. If not, here's a table with a mnemonic we created for you.

IPMAT	
Interphase	**I** is for **Interlude**
Prophase	**P** is for **Prepare**
Metaphase	**M** is for **Middle**
Anaphase	**A** is for **Apart**
Telophase	**T** is for **Two**

Purpose of Mitosis

Mitosis has two purposes:

- to produce daughter cells that are identical copies of the parent cell
- to maintain the proper number of chromosomes from generation to generation

For our purposes, we can say that mitosis occurs in just about every cell except sex cells. When you think of mitosis, remember: "Like begets like." Hair cells "beget" other hair cells, skin cells "beget" other skin cells, and so on. Mitosis is involved in growth, repair, and asexual reproduction.

HAPLOIDS VERSUS DIPLOIDS

Every organism has a certain number of unique chromosomes. For example, fruit flies have 4 chromosomes, humans have 23 chromosomes, and dogs have 39 chromosomes. It turns out that most eukaryotic cells in fact have two full sets of chromosomes—one set from each parent. Humans, for example, have two sets of 23 chromosomes, giving us our grand total of 46.

A cell that has two sets of chromosomes is a **diploid cell**, and the zygotic chromosome number is given as "$2n$." That means we have two copies of each chromosome.

If a cell has only one set of chromosomes, we call it a **haploid cell**. This kind of cell is given the symbol n. For example, we would say that the haploid number of chromosomes for humans is 23.

Remember that the n stands for the number of unique chromosomes.

Remember:

- *Diploid* refers to any cell that has two sets of chromosomes.
- *Haploid* refers to any cell that has one set of chromosomes.

Why do we need to know the terms *haploid* and *diploid*? Because they are extremely important when it comes to sexual reproduction. As we've seen, 46 is the normal diploid number for human beings, but there are only 23 different chromosomes. The duplicate versions of each chromosome are called **homologous chromosomes**. The homologous chromosomes that make up each pair are similar in size and shape and express those same traits (although, they might have different versions of those traits). This is the case in all sexually reproducing organisms. In fact, this is the essence of sexual reproduction: each parent donates half its chromosomes to its offspring.

Gametes

Although most cells in the human body are diploid, there are special cells that are haploid. These haploid cells are called **sex cells**, or **gametes**. Why do we have haploid cells?

As we've said, an offspring has one set of chromosomes from each of its parents. A parent, therefore, contributes a gamete with one set that will be paired with the set from the other parent to produce a new diploid cell, or zygote. This produces offspring that are a combination of their parents.

AN OVERVIEW OF MEIOSIS

Meiosis is the production of gametes. Since sexually reproducing organisms need only haploid cells for reproduction, meiosis is limited to sex cells in special sex organs called **gonads**. In males, the gonads are the **testes**, while in females they are the **ovaries**. The special cells in these organs—also known as **germ cells**—produce haploid cells (n), which then combine to restore the diploid ($2n$) number during fertilization.

$$\text{female gamete } (n) + \text{male gamete } (n) = \text{zygote } (2n)$$

When it comes to genetic variation, meiosis is a big plus. Variation, in fact, is the driving force of evolution. The more variation there is in a population, the more likely it is that some members of the population will survive extreme changes in the environment. Meiosis is far more likely to produce these sorts of variations than is mitosis, which therefore confers selective advantage on sexually reproducing organisms. We'll come back to this theme in Chapter 10.

A Closer Look at Meiosis

Meiosis actually involves two rounds of cell division: **meiosis I** and **meiosis II**.

Before meiosis begins, the diploid cell goes through interphase. Just as in mitosis, double-stranded chromosomes are formed during S phase.

Meiosis I

Meiosis I consists of four stages: prophase I, metaphase I, anaphase I, and telophase I.

Prophase I

Prophase I is a little more complicated than regular prophase. As in mitosis, the nuclear membrane disappears, the chromosomes become visible, and the centrioles move to opposite poles of the nucleus. But that's where the similarity ends.

The major difference involves the movement of the chromosomes. In meiosis, the chromosomes line up side-by-side with their counterparts (homologs). This event is known as **synapsis**.

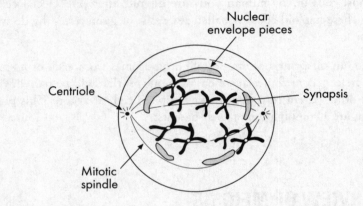

Synapsis involves two sets of chromosomes that come together to form a **tetrad** (or a **bivalent**). A tetrad consists of four chromatids. Synapsis is followed by **crossing-over**, the exchange of segments between homologous chromosomes.

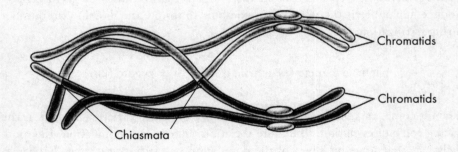

What's unique in prophase I is that pieces of chromosomes are exchanged between the homologous partners. This is one of the ways organisms produce genetic variation. By the end of prophase I, the chromosomes will have exchanged regions containing several **alleles**, or different forms of the same gene. This means that each chromosome might have a different mix of alleles than it began with. Remember, the two homologous chromosomes are the versions that came from the person's mother and father. So, a chromosome that has experienced crossing-over will now be a mix of alleles from the mother and father versions.

Metaphase I

As in mitosis, the chromosome pairs—now called tetrads—line up at the metaphase plate. By contrast, you'll recall that in regular metaphase, the chromosomes line up individually. One important concept to note is that the alignment during metaphase is random, so the copy of each chromosome that ends up in a daughter cell is random. Therefore, the gamete will be created with a mixture of genetic information from the person's father and mother. This means that the offspring created would be a combination of all four grandparents.

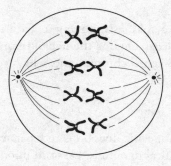

Anaphase I

During anaphase I, each pair of chromatids within a tetrad separates and moves to opposite poles. Notice that the chromatids do not separate at the centromere. The homologs separate with their centromeres intact.

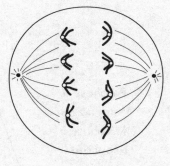

The chromosomes now go on to their respective poles.

Telophase I

During telophase I, the nuclear membrane forms around each set of chromosomes.

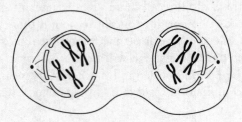

Finally, the cells undergo cytokinesis, leaving us with two daughter cells. Notice that at this point the nucleus contains the haploid number of chromosomes, but each chromosome is a duplicated chromosome, which consists of two chromatids.

Meiosis II

The purpose of the second meiotic division is to separate the sister chromatids, and the process is virtually identical to mitosis. Let's run through the steps in meiosis II.

During prophase II, the chromosomes once again condense and become visible. In metaphase II, the chromosomes move toward the metaphase plate. This time they line up single file, not as pairs. During anaphase II, the chromatids of each chromosome split at the centromere, and each chromatid is pulled to opposite ends of the cell. At telophase II, a nuclear membrane forms around each set of chromosomes and a total of *four* haploid cells are produced.

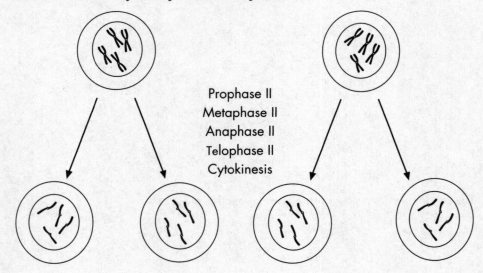

Prophase II
Metaphase II
Anaphase II
Telophase II
Cytokinesis

Gametogenesis

Meiosis is also known as **gametogenesis.** If sperm cells are produced, then meiosis is called **spermatogenesis**. During spermatogenesis, four sperm cells are produced for each diploid cell. If an egg cell or an ovum is produced, this process is called **oogenesis**.

Oogenesis is a little different from spermatogenesis. Oogenesis produces only one ovum, not four. The other three cells, called **polar bodies,** get only a tiny amount of cytoplasm and eventually degenerate. Why does oogenesis produce only one ovum? Because the female wants to conserve as much cytoplasm as possible for the surviving gamete, the **ovum**.

Here's a summary of the major differences between mitosis and meiosis.

MITOSIS	MEIOSIS
• Occurs in somatic (body) cells	• Occurs in germ (sex) cells
• Produces identical cells	• Produces gametes
• Diploid cell → diploid cells	• Diploid cell → haploid cells
• 1 cell becomes 2 cells	• 1 cell becomes 4 cells
• Number of divisions: 1	• Number of divisions: 2

Meiotic Errors

Sometimes, a set of chromosomes has an extra or a missing chromosome. This occurs because of **nondisjunction**—the chromosomes failed to separate properly during meiosis. This error, which produces the wrong number of chromosomes in a cell, usually results in miscarriage or significant genetic defects. For example, humans typically have two copies of each chromosome, but individuals with **Down syndrome** have three—instead of two—copies of the 21st chromosome.

Chromosomal abnormalities also occur if one or more segments of a chromosome break and are either lost or reattach to another chromosome. The most common example is **translocation** (a segment of a chromosome moves to another chromosome).

Here's an example of a translocation.

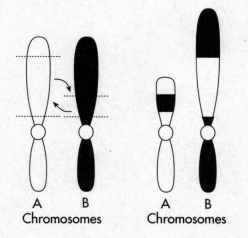

A   B
Chromosomes

A   B
Chromosomes

Fortunately, in most cases, damaged DNA can usually be repaired by special repair enzymes.

KEY TERMS

cell division

G_0 phase

cell cycle

interphase

mitosis

G_1 phase

S phase

G_2 phase

sister chromatids

centromere

cyclins

cyclin-dependent kinases (CDKs)

cell cycle checkpoints

apoptosis

cancer

oncogene

tumor suppressor gene

prophase

metaphase

anaphase

telophase

chromatin

spindle fibers

kinetochores

metaphase plate

cytokinesis

cleavage furrow

cell plate

diploid cell

haploid cell

homologous chromosomes

sex cells (or gametes)

meiosis

gonads

testes

ovaries

germ cells

meiosis I

meiosis II

synapsis

tetrad (or bivalent)

crossing-over (or recombination)

alleles

gametogenesis

spermatogenesis

oogenesis

polar bodies

ovum

nondisjunction

Down syndrome

translocation

Summary

- The cell cycle is divided into interphase and mitosis, or cellular division.

- The three stages of interphase are G_1, G_2, and S phase.
    - S phase is the "synthesis" phase, when chromosomes replicate.
    - Growth and preparation for mitosis occur in G_1 and G_2.

- Cell cycle progression is controlled by checkpoint pathways and CDK/cyclin complexes.

- Cancer occurs when cells grow abnormally and spread to other parts of the body. Tumor-suppressor genes are genes that prevent the cell from dividing when it shouldn't. Proto-oncogenes are genes that help the cell divide. If either type is mutated (a mutated proto-oncogene is called an oncogene), it can lead to cell growth that is out of control.

- Mitosis is cellular division and occurs in four stages: prophase, metaphase, anaphase, and telophase.
    - During prophase, the nuclear envelope disappears and chromosomes condense.
    - Next is metaphase, when chromosomes align at the metaphase plate and mitotic spindles attach to kinetochores.
    - Anaphase pulls the chromosomes away from the center.
    - Telophase terminates mitosis, and the two new nuclei form.
    - The process of cytokinesis, which occurs during telophase, ends mitosis, as the cytoplasm and plasma membranes pinch to form two distinct, identical daughter cells.

- Meiosis produces four genetically distinct haploid gametes (sperm or egg cells). Meiosis involves two rounds of cell division: meiosis I and meiosis II.
    - Meiosis I is when the homologous chromosome pairs separate, but before they can they undergo crossing-over and swap some DNA.
    - During meiosis II, the sister chromatids separate.
    - Mutations, such as nondisjunction events or whole chromosome translocations, can occur as a result of crossing-over.

Chapter 8 Drill

Answers and explanations can be found in Chapter 16.

1. A scientist is testing new chemicals designed to stop the cell cycle at various stages of mitosis. Upon applying one of the chemicals, she notices that all of the cells appear as shown below. Which of the following best explains how the chemical is likely acting on the cells?

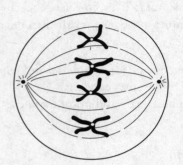

(A) The chemical has arrested the cells in prophase and has prevented attachment of the spindle fibers to the kinetochore.

(B) The chemical has arrested the cells in metaphase and has prevented dissociation of the spindle fibers from the centromere.

(C) The chemical has arrested the cells in metaphase and is preventing the shortening of the spindle fibers.

(D) The chemical has arrested the cells in anaphase and is preventing the formation of a cleavage furrow.

Questions 2 and 3 refer to the following graph and paragraph.

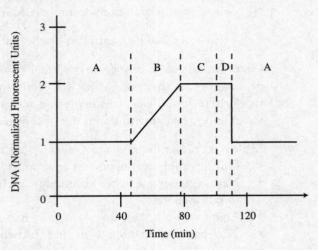

An experiment is performed to evaluate the amount of DNA present during a complete cell cycle. All of the cells were synced prior to the start of the experiment. During the experiment, a fluorescent chemical was applied to cells, which would fluoresce only when bound to DNA. The results of the experiment are shown above. Differences in cell appearance by microscopy or changes in detected DNA were determined to be phases of the cell cycle and are labeled with the letters A–D.

2. Approximately how long does S phase take to occur in these cells?

    (A) 15 min
    (B) 20 min
    (C) 30 min
    (D) 40 min

3. During which of the labeled phases of the experiment would the cell undergo anaphase?

    (A) Phase A
    (B) Phase B
    (C) Phase C
    (D) Phase D

Question 4 represents a question requiring a numeric answer. Calculate the correct answer for each question.

4. A new mammal has recently been discovered in the Amazonian jungle. A karyotype was performed on gametic cells and revealed that the animal had 13 completely unique chromosomes. How many homologous pairs would you expect to find in a diploid cell of the organism following completion of S phase? Give your answer to the nearest whole number.

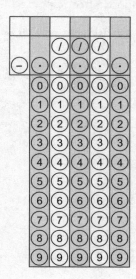

5. Trisomy 21, which results in Down syndrome, results from nondisjunction of chromosome 21 in humans. Nondisjunction occurs when two homologous chromosomes, or two sister chromatids, do not separate. Which of the following describes the mechanism of this defect?

    (A) During DNA replication in S phase of the cell cycle, the two new strands do not separate.
    (B) During mitosis, at the metaphase plate, non-sister chromatids do not separate.
    (C) The mitotic spindle attaches to chiasmata rather than kinetochores.
    (D) The same microtubule in the spindle attaches to both sister chromatids during meiosis II.

6. A researcher isolates DNA from different types of cells and determines the amount of DNA for each type of cell. The samples may contain cells in various stages of the cell cycle. In order of increasing DNA content, which of the following would have the least amount of DNA to the greatest amount of DNA?

    (A) Sperm, neuron, liver
    (B) Ovum, muscle, taste bud
    (C) Liver, heart, intestine
    (D) Skin, neuron, ovum

7. Which of the following mutations would be least likely to have any discernable phenotype on the individual?

    (A) Translocation of the last 1,000 base pairs of chromosome 1 onto chromosome 2
    (B) Nondisjunction of chromosome 7 to produce trisomy-7
    (C) Nondisjunction of chromosome 8 to produce monosomy-8
    (D) Deletion of 400 base pairs, resulting in the loss of an enhancer

REFLECT

Respond to the following questions:

- Which topics from this chapter do you feel you have mastered?

- Which content topics from this chapter do you feel you need to study more before you can answer multiple-choice questions correctly?

- Which content topics from this chapter do you feel you need to study more before you can effectively compose a free response?

- Was there any content that you need to ask your teacher or another person about?

Chapter 9
Heredity

GREGOR MENDEL: THE FATHER OF GENETICS

What is genetics? In its simplest form, genetics is the study of heredity. It explains how certain characteristics are passed on from parents to children. Much of what we know about genetics was discovered by the monk **Gregor Mendel** in the 19th century. Since then, the field of genetics has vastly expanded.

Let's begin then with some of the fundamental points of genetics:

- **Traits**—or expressed characteristics—are influenced by one or more of your genes. Remember, a **gene** is simply a chunk of DNA that codes for a particular recipe. Different recipes affect different traits. DNA is passed from generation to generation, and from this process the genes and the traits associated with those genes are inherited. Within a chromosome, there are many genes, each controlling the inheritance of a particular trait. For example, in pea plants, there's a gene on the chromosome that codes for seed coat. The position of a gene on a chromosome is called a **locus**.

- Diploid organisms (organisms that have two sets of chromosomes) usually have two copies of a gene, one on each **homologous chromosome**. Homologous chromosomes are the two copies/versions of a chromosome that diploids have. Humans have 23 pairs of homologous chromosomes.

  These copies may be different from one another—that is, they may be **alleles**, or alternate versions/flavors of the same gene. For example, if we're talking about the height of a pea plant, there's an allele for tall and an allele for short. In other words, both alleles are alternate forms of the gene for height.

- When an organism has two identical alleles for a given trait, the organism is **homozygous**. If an organism has two *different* alleles for a given trait, the organism is **heterozygous**.

- When discussing the physical appearance of an organism, we refer to its **phenotype**. The phenotype tells us what the organism looks like. When talking about the genetic makeup of an organism, we refer to its **genotype**. The genotype tells us which alleles the organism possesses.

- An allele can be **dominant** or **recessive**. This is determined by which allele "wins out" over the other in a heterozygote. The convention is to assign one of two letters for the two different alleles. The dominant allele receives a capital letter and the recessive allele receives a lowercase of the same letter. For instance, we might give the dominant allele for height in pea plants a T for tall. This means that the recessive allele would be t for short.

Name	Genotype	Phenotype
homozygous dominant	TT	Tall
homozygous recessive	tt	Short
heterozygous	Tt	Tall

One of the major ways the testing board likes to test genetic information is by having you do crosses. Crosses involve the mating of hypothetical organisms with specific phenotypes and genotypes. We'll look at some examples in a moment, but for now, keep these test-taking tips in mind:

- Label each generation in the cross. The first generation in an experiment is always called the **parent**, or **P1 generation**. The offspring of the P1 generation are called the **filial**, or **F1 generation**. Members of the next generation, the grandchildren, are called the **F2 generation**.
- Always write down the symbols you're using for the alleles, along with a key to remind yourself what the symbols refer to. Use uppercase for dominant alleles and lowercase for recessive alleles.

Now let's look at some basic genetic principles.

MENDELIAN GENETICS

One of Mendel's hobbies was to study the effects of cross-breeding on different strains of pea plants. Mendel worked exclusively with true-breeder pea plants. This means the plants he used were genetically pure and consistently produced the same traits. For example, tall plants always produced tall plants; short plants always produced short plants. Through his work, he came up with three principles of genetics: the **Law of Dominance**, the **Law of Segregation**, and the **Law of Independent Assortment**.

The Law of Dominance

Mendel crossed two true-breeding plants with contrasting traits: tall pea plants and short pea plants.

To his surprise, when Mendel mated these plants, the characteristics didn't blend to produce plants of average height. Instead, all the offspring were tall.

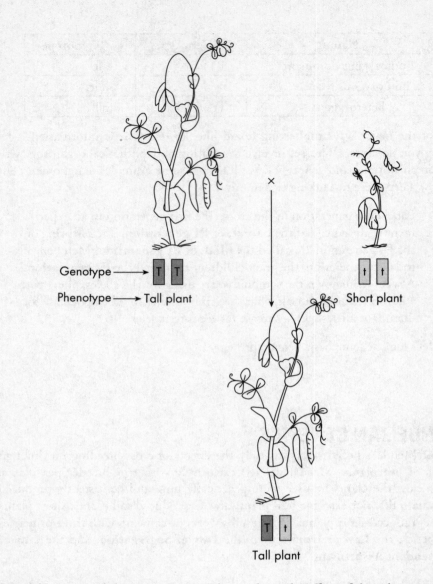

Genotype ⟶ T T
Phenotype ⟶ Tall plant

×

Short plant
t t

T t
Tall plant

Know Your Way Around a Square

A Punnett square is just a visual way to see all the possible offspring combos that could result from two parents mating. The possible gametes for each parent go on the sides and then they get matched up with each other to show the possible offspring genotypes.

Remember, these squares are looking only at one (or sometimes two) genes. So, they show possible offspring genotypes for that one gene only.

Mendel recognized that one trait must be masking the effect of the other trait. This is called the Law of Dominance. The dominant tall allele, T, somehow masked the presence of the recessive short allele, t. Consequently, all a plant needs is one tall allele to make it tall.

Monohybrid Cross

A **monohybrid cross** occurs when two individuals that are heterozygous for two traits are crossed. However, in order to be sure that you have heterozygotes, the first step is to cross true-breeding plants of allele #1 with true-breeding plants of allele #2. Follow along as these plants are crossed.

A simple way to represent a cross is to set up a **Punnett square**. Punnett squares are used to predict the results of a cross. Let's construct a Punnett square for the cross between Mendel's tall and short pea plants. Let's first designate the alleles for each plant. As we saw earlier, we can use the letter "T" for the tall dominant allele and "t" for the recessive short allele.

Since one parent was a pure, tall pea plant, we'll give it two dominant alleles (TT homozygous dominant). The other parent was a pure, short pea plant, so we'll give it two recessive alleles (tt homozygous recessive). To look at the types of offspring that these parents could produce we have to figure out the gametes that each parent can possibly make (and thus contribute to the offspring).

Each parent has two copies of the height gene and each gamete gets one copy. This means each parent can make two types of gametes. For the homozygous dominant parent, the two copies are T and T. This means they can make gametes with T or gametes with T. The homozygous recessive can make gametes with t and with t. We will put the possible gametes for each parent on the top and left side of the square as shown below.

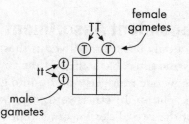

Now we can fill in the four boxes by matching the letters. What are the results for the F1 generation?

	T	T
t	Tt	Tt
t	Tt	Tt

Each offspring received one allele from each parent. They all received one T and one t. They're all Tt! So, even though the parents were homozygous, the offspring are heterozygous: they possess one copy of each allele.

The Law of Segregation

Next, Mendel took the offspring and self-pollinated them. Let's use a Punnett square to spell out the results. This time we're mating the offspring of the first generation—F1; we are crossing those heterozygotes. Take a look at the results:

	T	t
T	TT	Tt
t	Tt	tt

F2 Generation

Check out the lower right box. One of the offspring could be a short pea plant!

Although all of the F1 plants were to be tall, they were heterozygotes that could make gametes with just the short t allele. In other words, their T and t alleles were segregated (separated) during gamete formation. This is an example of the Law of Segregation. It just means that each gamete gets only one of the two copies of a gene. Gametes are haploid.

What about the genotype and phenotype for this cross? Remember, genotype refers to the genetic makeup of an organism, whereas phenotype refers to the appearance of the organism. Using the results of our Punnett square, what is the ratio of phenotypes and genotypes in the offspring?

Let's sum up the results. We have four offspring with two different phenotypes: three of the offspring are tall, whereas one of them is short. On the other hand, we have three genotypes: 1 TT, 2 Tt, and 1 tt.

Here's a summary of the results:

- The ratio of phenotypes is 3:1 (three tall:one short).
- The ratio of genotypes is 1:2:1 (one TT:two Tt:one tt).

The Law of Independent Assortment

So far, we have looked at only one trait: tall versus short. What happens when we study two traits at the same time? Each allele of the two traits will get segregated into two gametes, but how one trait gets split up into gametes has no bearing on how the other trait gets split up. In other words, the two alleles for each trait are sorted independently of the two alleles for other traits. For example, let's look at two traits in pea plants: height and color. When it comes to height, a pea plant can be either tall (T) or short (t). As for color, the plant can be either green (G) or yellow (g), with green being dominant. Each gene comes in two alleles, so this gives us four alleles total. By the Law of Independent Assortment, these four alleles can combine to give us four different gametes.

TG Tg tG tg

In other words, each allele of the height gene can be paired with either allele of the color gene. The reason alleles segregate independently is because chromosomes segregate independently. During meiosis I, each pair of homologous chromosomes is split, and which one aligns left or right during metaphase I is different for each pair.

Dihybrid Cross

Now that we know that different genes assort independently into gametes, let's look at a cross of two traits using those four gametes that we created above. A **dihybrid cross** is just like the monohybrid, but two heterozygotes for two genes are crossed.

Here is the Punnett Square for a cross between two double heterozygotes (Tt Gg).

	TG	Tg	tG	tg
TG	TTGG	TTGg	TtGG	TtGg
Tg	TTGg	TTgg	TtGg	Ttgg
tG	TtGG	TtGg	ttGG	ttGg
tg	TtGg	Ttgg	ttGg	ttgg

This is an example of the Law of Independent Assortment. Each of the traits segregated independently. Don't worry about the different combinations in the cross—you'll make yourself dizzy with all those letters. Simply memorize the phenotype ratio of the pea plants. For the 16 offspring there are:

- 9 tall and green
- 3 tall and yellow
- 3 short and green
- 1 short and yellow

That's 9:3:3:1, but don't forget the circumstances that this applies. It is only when two heterozygotes for two genes are crossed. It is not just a magic ratio that always works.

Rules of Probability

The Punnett square method works well for monohybrid crosses and helps us visualize the possible combinations. However, a better method for predicting the likelihood of certain results from a dihybrid cross is to apply the Law of Probability. The law states that the probability that two or more independent events will occur simultaneously is equal to the product of the probability that each will occur independently. To illustrate the product rule, let's consider again the cross between two identical dihybrid tall, green plants with the genotype TtGg.

> Probability of 2 traits occurring together = probability of trait A occurring × probability of trait B occurring

To find the probability of having a tall, yellow plant, simply multiply the probabilities of each event. If the probability of being tall is $\frac{3}{4}$ and the probability of being yellow is $\frac{1}{4}$, then the probability of being tall and yellow is $\frac{3}{4} \times \frac{1}{4} = \frac{3}{16}$.

Let's summarize Mendel's three laws.

Probability can be expressed as a fraction, percentage, or a decimal. Remember that this rule works only if the results of one cross are not affected by the results of another cross.

SUMMARY OF MENDEL'S LAWS	
Laws	**Definition**
Law of Dominance	One trait masks the effects of another trait.
Law of Segregation	Each gamete gets only one of the copies of each gene.
Law of Independent Assortment	Each pair of homologous chromosomes splits independently, so the alleles of different genes can mix and match.

Test Cross

Suppose we want to know if a tall plant is homozygous (TT) or heterozygous (Tt). Its physical appearance doesn't necessarily tell us about its genetic makeup. The only way to determine its genotype is to cross the plant with a recessive, short plant, tt. This is known as a **test cross**. Using the recessive plant, there are only two possibilities of what is being crossed. Our mystery tall plant could be TT or Tt so when we mate it to a short plant, these are the possible situations: (1) TT × tt or (2) Tt × tt. Let's take a look.

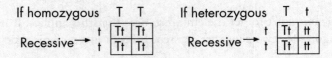

If none of the offspring is short, our original plant must have been homozygous, TT. If, however, even one short plant appears in the bunch, we know that our original pea plant was heterozygous, Tt. In other words, it wasn't a pure-breeding plant. A test cross uses a recessive organism to determine the genotype of an organism of unknown genotype.

Linked genes

Sometimes genes on the *same* chromosome stay together during assortment and move as a group. The group of genes is considered linked and tends to be inherited together. For example, the genes for flower color and pollen shape are linked on the same chromosomes and show up together.

Since **linked genes** are found on the same chromosome, they cannot segregate independently. This violates the Law of Independent Assortment.

Let's pretend that the height and color genes we looked at in the dihybrid cross are linked. A heterozygote for both traits still has two alleles for height (T and t) and two alleles for color (G and g). However, because height and color are located on the same chromosome, the allele for height and the allele for color are physically linked. For example, maybe the heterozygote has one chromsome with Tg and one chromosome with tG. When gametes are formed, the T and g will travel together, and the t and G will travel together and be packaged into a gamete together. So, in the unlinked dihybrid shown earlier there were four possible gamete combinations (TG, Tg, tG, tg), but now there are only two (Tg and tG). The only way to physically separate linked alleles is by crossing-over. If a crossover event occurs between the linked genes, then recombinant gametes can occur.

If the genes were unlinked, then the the four gametes (TG, Tg, tG, tg) would be equally likely (as we saw when we talked about independent assortment). However, if certain combinations of alleles are found more often in offspring (are likely appearing more often in gametes), then this is a sign of possible linkage.

Look at the example below from a cross of TtGg and ttgg. If you draw out the Punnett square, you should get equal numbers of each phenotype occurring if the genes are assumed to be unlinked. Yet, what if the numbers below were the actual numbers of offspring of each phenotype?

Tall and Green: 6 offspring

short and Green: 37 offspring

Tall and yellow: 45 offspring

short and yellow: 3 offspring

There are more Green/short and yellow/Tall offspring. This is a sign that these alleles are linked on a chromosome because they occur more often than they statistically should. This shows that most of the gametes of the TtGg parent had tG together and Tg together since those gametes would result in Green/short and yellow/Tall offspring.

So, where did the Green/Tall and yellow/short plants come from? They are the result of gametes from the TtGg parent that are TG and tg, respectively. If the Tg and tG alleles were linked in that parent, how could they produce a few TG and tg gametes? The answer is through recombination! Even though those alleles were not found on the same chromosome, recombination occurred during meiosis which swapped them. This allowed gametes to be formed with TG and tg, and offspring to occur that were Tall/Green and short/yellow. The offspring formed from reccombination events are called **recombinants**. The **percentage of recombination** (or **recombination frequency**) can be determined by adding up the recombinants and dividing by the total number of offspring.

$$9/91 = 9.9\%$$

This percentage can also be used as a measure of how far apart the genes are. The distance on a chromosome is measured in **map units** or centimorgans.

The frequency of crossing-over between any two linked alleles is proportional to the distance between them. That is, the farther apart two linked alleles are on a chromosome, the more often the chromosome will cross over between them. This finding led to recombination mapping—mapping of linkage groups with each map unit being equal to 1 percent recombination. For example, if two linked genes, A and B, recombine with a frequency of 15 percent, and B and C recombine with a frequency of 9 percent, and A and C recombine with a frequency of 24 percent, what is the sequence and the distance between them?

The sequence and the distance of A-B-C is

15 units   9 units
A_____B_____C

If the recombination frequency between A and C had been 6 percent instead of 24 percent, the sequence and distance of A-B-C would instead be

6 units   9 units
A_____C_____B

SEX-LINKED TRAITS

We already know that humans contain 23 pairs of chromosomes. Twenty-two of the pairs of chromosomes are called **autosomes**. They code for many different traits. The other pair contains the **sex chromosomes**. This pair determines the sex of an individual. A female has two X chromosomes. A male has one X and one Y chromosome—an X from his mother and a Y from his father. Some traits, such as **color blindness** and **hemophilia**, are carried on sex chromosomes. These are called **sex-linked traits**. Most sex-linked traits are found on the X chromosome and are more properly referred to as "X-linked."

Since males have one X and one Y chromosome, what happens if a male has an X-chromosome with the color blindness allele? Unfortunately, he'll express the sex-linked trait, even if it is recessive. Why? Because his one and only X chromosome is color blind-X. He doesn't have another X to mask the effect of the color blind-X. However, if a female has only one color blind-X chromosome, she won't express a recessive sex-linked trait. For her to express the trait, she has to inherit two color blind-X chromosomes. A female with one color blind-X is called a **carrier**. Although she does not exhibit the trait, she can still pass it on to her children.

You can also use the Punnett square to figure out the results of sex-linked traits. Here's a classic example: a male who has normal color vision and a woman who is a carrier for color blindness have children. How many of the children will be color-blind? To figure out the answer, let's set up a Punnett square.

$$
\begin{array}{c|c|c}
 & \text{X} & \bar{\text{X}} \leftarrow \text{Mother} \\
\hline
\text{X} & \text{XX} & \overline{\text{XX}} \\
\hline
\text{Y} & \text{XY} & \overline{\text{X}}\text{Y} \\
\end{array}
$$

Father →

$\bar{\text{X}}$ = color blind-X

Notice that we placed a bar above any color blind-X to indicate the presence of a color blindness allele. And now for the results. The couple could have a son who is color-blind, a son who sees color normally, a daughter who is a carrier, or a daughter who sees color normally.

Barr Bodies

A look at the cell nucleus of normal females will reveal a dark-staining body known as a **Barr body**. A Barr body is an X chromosome that is condensed and visible. In every female cell, one X chromosome is activated and the other X chromosome is deactivated during embryonic development. Surprisingly, the X chromosome destined to be inactivated is randomly chosen in each cell. Therefore, in every tissue in the adult female, one X chromosome remains condensed and inactive. However, this X chromosome is replicated and passed on to a daughter cell. X-inactivation is why it is okay that females have two X chromosomes and males have only one. After X-inactivation, it is like everyone has one copy.

Other Inheritance Patterns

Not all patterns of inheritance obey the principles of Mendelian genetics. In fact, many traits we observe are due to a combined expression of alleles. Here are a couple of examples of non-Mendelian forms of inheritance:

- **Incomplete dominance (blending inheritance):** In some cases, the traits will blend. For example, if you cross a white snapdragon plant (genotype WW) with a red snapdragon plant (RR), the resulting progeny will be pink (RW). In other words, neither color is dominant over the other.
- **Codominance:** Sometimes you'll see an equal expression of both alleles. For example, an individual can have an AB blood type. In this case, each allele is equally expressed. That is, both the A allele and the B allele are expressed ($I^A I^B$). That's why the person is said to have AB blood. The expression of one allele does not prevent the expression of the other.
- **Polygenic inheritance:** In some cases, a trait results from the interaction of *many* genes. Each gene will have a small effect on a particular trait. Height, skin color, and weight are all examples of polygenic traits.
- **Non-nuclear inheritance**: Apart from the genetic material held in the nucleus, there is also genetic material that is contained in the mitochondria. The mitochondria are always provided by the egg during sexual reproduction, so mitochondrial inheritance is always through the maternal line, not the male line.

Other Sex-Linked Diseases

- Duchenne muscular dystrophy
- Vitamin D resistant rickets
- Hereditary nephritis
- Juvenile gout

KEY TERMS

Gregor Mendel
trait
genes
locus
homologous chromosomes
alleles
homozygous
heterozygous
phenotype
genotype
dominant
recessive
parent, or P1 generation
filial, or F1 generation
F2 generation
law of dominance
law of segregation
law of independent assortment
monohybrid cross

Punnett square
dihybrid cross
test cross
linked genes
recombinants
percentage of recombination
  (recombination frequency)
map units
autosomes
sex chromosomes
color blindness
hemophilia
sex-linked traits
carrier
Barr body
incomplete dominance
codominance
polygenetic inheritance
non-nuclear inheritance

Summary

- Genetics is the study of heredity. The foundation of genetics is that traits are produced by genes, which are found on chromosomes.

- Diploid organisms carry two copies of every gene, one on each of two homologous chromosomes, one that came from their mother and one that came from their father.

- When the two alleles are the same, the organism is homozygous for that trait. When the two alleles are different, the organism is heterozygous for the trait.

- The genes are described as the genotype of the organism. The physical expression of the traits, or what you see in the organism, is the phenotype.

- Homozygous individuals have only one allele that will determine the phenotype. However, heterozygous individuals have two different alleles that could contribute to the phenotype. The allele that "wins out" and directs the phenotype is the dominant allele. The one that doesn't show up in the phenotype is the recessive allele.

- Simple genetics with dominants and recessives was illuminated by Gregor Mendel. In Mendelian genetics, there are three important laws: the Law of Dominance, the Law of Segregation, and the Law of Independent Assortment.

- If two genes are on the same chromosome they are linked. The alleles on the same chromosome are locked together and travel together during gamete formation. Recombination frequency can be used to measure how far apart two genes are on a chromosome.

- Humans have 23 pairs of homologous chromosomes for a total of 46. Of these, 22 pairs are autosomes and 2 are designated as sex chromosomes:
  - Females have two X chromosomes; males have one X and one Y chromosome.
  - Because females have two X chromosomes, one is inactivated in each cell and condenses into a Barr body.

- Non-Mendelian genetics refers to a situation in which traits do not follow the Mendelian laws. Examples include the following:
  - incomplete dominance
  - codominance
  - polygenic inheritance

Chapter 9 Drill

Answers and explanations can be found in Chapter 16.

1. Two expecting parents wish to determine all possible blood types of their unborn child. If both parents have an AB blood type, which of the following blood types will their child NOT possess?

   (A) A
   (B) B
   (C) AB
   (D) O

2. A new species of tulip was recently discovered. A population of pure red tulips was crossed with a population of pure blue tulips. The resulting F1 generation was all purple. This result is an example of which of the following?

   (A) Complete dominance
   (B) Incomplete dominance
   (C) Codominance
   (D) Linkage

3. In pea plants, tall (T) is dominant over short (t) and green (G) is dominant over yellow (g). If a pea plant that is heterozygous for both traits is crossed with a plant that is recessive for both traits, approximately what percentage of the progeny plants will be tall and yellow?

   (A) 0%
   (B) 25%
   (C) 66%
   (D) 75%

Questions 4-6 refer to the following pedigree tree and paragraph.

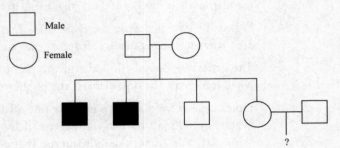

Male

Female

?

Hemophilia is an X-linked disease associated with the inability to produce specific proteins in the blood-clotting pathway. Shown above is a family pedigree tree in which family members afflicted with the disease are shown with filled-in squares (male) or circles (females). A couple is trying to determine the likelihood of passing on the disease to their future children (represented by the ? symbol above) because the hemophilia runs in the woman's family.

4. Assuming that the woman in the couple is a carrier, what is the probability that the couple's first son will have hemophilia?

   (A) 0%
   (B) 25%
   (C) 50%
   (D) 100%

5. Why were both women in the family tree above free of the disease?

   (A) They were lucky because they didn't receive an X chromosome with the diseased allele.
   (B) They cannot get hemophilia because it is associated only with a diseased Y chromosome.
   (C) They are carriers and will get the disease only if they have children.
   (D) They have at least one normal (unaffected) copy of the X chromosome.

6. Turner syndrome is a disease in which an individual is born with only a single X chromosome. Suppose the woman in the couple is a carrier for hemophilia and has a child with Turner syndrome. Would this child have the disease?

   (A) Yes, because the child would have only one copy of the X chromosome and it would be affected.
   (B) No, because women cannot be affected by hemophilia.
   (C) No, because the child would have to receive a normal copy of the X chromosome from its mother.
   (D) Maybe, it depends on which X chromosome she receives from her mother.

7. Neuronal ceroid lipofuscinosis (NCL) is a group of autosomal recessive diseases characterized by blindness, loss of cognitive and motor function, and early death. One of the genes that is mutated in this disease is CLN3. When functional CLN3 protein is absent, neurons die because of increased storage material in the cells, presumably because the lysosomes aren't working properly. What is a valid explanation as to why parents of children with this devastating disease can be completely normal?

   (A) Parents of the autosomal recessive disease must both be carriers of the mutation on the CLN3 gene.
   (B) Carriers of the CLN3 mutation genotype do not show the phenotype because one normal CLN3 allele is present to provide a functioning CLN3 protein.
   (C) Redundant proteins take over the function of the mutant CLN3 protein in the parents.
   (D) X inactivation prevents expression of the CLN3 mutated protein in the parents.

8. A breeder of black Labrador puppies notices that in his line a gene predisposing his dogs for white spots has arisen. He believes the white spot allele is autosomal recessive, and he wants to prevent it from continuing in his dogs. He sees that one of his female dogs, Speckle, has white spots, and he therefore assumes she is homozygous for the gene. But he must determine which males in his pack are heterozygous to avoid breeding them in the future. What is a reasonable plan?

   (A) Stop breeding Speckle.
   (B) Pair Speckle with a male with white spots.
   (C) Perform a testcross of Speckle with a black Labrador male. If any pups are born with white spots, stop breeding that male because he is heterozygous.
   (D) Breed Speckle with a chocolate Labrador.

9. Which of the following is a reason why certain traits do not follow Mendel's Law of Independent Assortment?

   (A) The genes are linked on the same chromosome.
   (B) It applies only to eukaryotes.
   (C) Certain traits are not completely dominant.
   (D) Heterozygotes have both alleles

10. The gene for coat color in some breeds of cats is found on the X chromosome. Calico cats are mottled with orange and black colorings. What is a possible explanation for the fact that true calico cats are only female?

    (A) The allele for coat color is randomly chosen by X-inactivation.
    (B) The coat color is linked with genes that are lethal to male cats, so male calico cats never live past birth.
    (C) Coat orange and black color is expressed only in female cats; males have only one color.
    (D) Male cats have Y chromosomes.

REFLECT

Respond to the following questions:

- Which topics from this chapter do you feel you have mastered?

- Which content topics from this chapter do you feel you need to study more before you can answer multiple-choice questions correctly?

- Which content topics from this chapter do you feel you need to study more before you can effectively compose a free response?

- Was there any content that you need to ask your teacher or another person about?

Chapter 10
Evolutionary
Biology

All of the organisms we see today arose from earlier organisms. This process, known as **evolution**, can be described as a change in a population over time. Interestingly, however, the driving force of evolution, called **natural selection**, operates on the level of the individual. In other words, evolution is defined in terms of populations but occurs in terms of individuals.

NATURAL SELECTION

What is the basis of our knowledge of evolution? Much of what we now know about evolution is based on the work of **Charles Darwin**. Darwin was a 19th-century British naturalist who sailed the world in a ship named the HMS *Beagle*. Darwin developed his theory of evolution based on natural selection after studying animals in the Galápagos Islands and other places.

He observed that there were similar animals on the various islands, but the beaks varied in length on the finches, and the necks varied in length on the tortoises. There must be a reason why animals in different areas had different traits. Darwin concluded that it was impossible for the finches and tortoises of the Galápagos simply to "grow" longer beaks or necks as needed. Rather, those traits were inherited and passed on from generation to generation. So, he decided there must be a way for populations to evolve and change their traits (i.e., a population of finches developing longer beaks).

He decided that there must have been a variety of beak lengths originally, but only the longest one was particularly helpful. Since those finches could eat better, survive better, and reproduce better, they were more likely to contribute offspring to the next generation. Over many years, the long-beak finches became more and more plentiful with each generation until long beaks were the "norm." In another example, on the first island Darwin studied, there must once have been short-necked tortoises. Unable to reach the higher vegetation, these tortoises eventually died off, leaving only those tortoises with longer necks. Consequently, evolution has come to be thought of as "the survival of the fittest": only those organisms most fit to survive will survive and reproduce.

Darwin elaborated his theory in a book entitled *On the Origin of Species*. In a nutshell, here's what Darwin observed:

- Each species produces more offspring than can survive.
- These offspring compete with one another for the limited resources available to them.
- Organisms in every population vary.
- The fittest offspring, or those with the most favorable traits, are the most likely to survive and therefore produce a second generation.

Lamarck and the Long Necks

Darwin was not the first to propose a theory explaining the variety of life on Earth. One of the most widely accepted theories of evolution in Darwin's day was that proposed by **Jean-Baptiste de Lamarck**.

In the 18th century, Lamarck had proposed that *acquired* traits were inherited and passed on to offspring. For example, in the case of Lamarck's giraffes, Lamarck's theory said that the giraffes had long necks because they were constantly reaching for higher leaves while feeding. This theory is referred to as the "law of use and disuse," or, as we might say now, "use it or lose it." According to Lamarck, giraffes have long necks because they constantly *use* them.

We know now that Lamarck's theory was wrong: acquired changes—that is, changes at a "macro" level in an organism's regular (somatic) body cells—will not appear in the gamete cells. For example, if you were to lose one of your fingers, your children would not inherit this trait. The gametes that your body makes include copies of your regular old genome, not a new version of a genome that is determined by how careful you are with a table saw.

Evidence for Evolution

In essence, nature "selects" which living things survive and reproduce. Today, we find support for the theory of evolution in several areas:

- **Paleontology**, or the study of fossils: paleontology has revealed to us both the great variety of organisms (most of which, including trilobites, dinosaurs, and the woolly mammoth, have died off) and the major lines of evolution.

- **Biogeography**, or the study of the distribution of **flora** (plants) and **fauna** (animals) in the environment: scientists have found related species in widely separated regions of the world. For example, Darwin observed that animals in the Galápagos have traits similar to those of animals on the mainland of South America. One possible explanation for these similarities is a common ancestor. As we'll see below, there are other explanations for similar traits. However, when organisms share multiple traits, it's pretty safe to say that they also shared a common ancestor.

- **Embryology**, or the study of the development of an organism: if you look at the early stages in vertebrate development, all the embryos look alike! For example, all vertebrates—including fish, amphibians, birds, and even humans—show fishlike features called gill slits.

- **Comparative anatomy**, or the study of the anatomy of various animals: scientists have discovered that some animals have similar structures that serve different functions. For example, a human's arm, a dog's leg, a bird's wing, and a whale's fin are all the same appendages, though they have evolved to serve different purposes. These structures, called **homologous structures**, also point to a common ancestor.

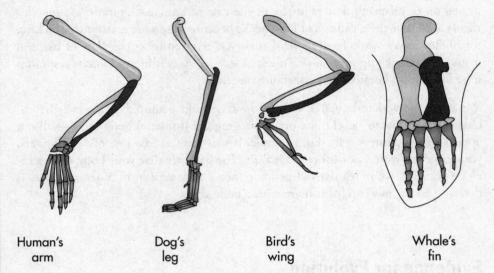

Human's arm   Dog's leg   Bird's wing   Whale's fin

In contrast, sometimes animals have features with the same function but that are structurally different. A bat's wing and an insect's wing, for example, are both used to fly. They therefore have the same function, but they have evolved totally independently of one another. These are called **analogous structures**. Another classic example of an analogous structure is the eye. Though scallops, insects, and humans all have eyes, these three different types of eyes are thought to have evolved entirely independently of one another. They are therefore analogous structures.

- **Molecular biology:** Perhaps the most compelling proof of all is the similarity at the molecular level. Today, scientists can examine the nucleotide and amino acid sequences of different organisms. From these analyses, we've discovered that organisms that are closely related have a greater proportion of sequences in common than distantly related species. For example, humans don't look much like chimpanzees. However, by some estimates, as much as 99 percent of our genetic code is identical to that of a chimp.

Across every domain, there are certain universal molecules and structures.

COMMON ANCESTRY

By using the evidence mentioned above, scientists can get a good handle on how evolution of certain species occurred. It all comes down to who has common ancestors. Life started somewhere. Some original life-form is the **common ancestor** to all life, but where things went from there can become quite convoluted.

Scientists use charts called **phylogenetic trees,** or cladograms, to study the relationships between organisms. They always begin with the common ancestor and then branch out. Anytime there is a fork in the road, it is called a common ancestor node. These common ancestors likely do not exist anymore, but they are the point at which evolution went in two directions. One direction eventually led to one species, and the other eventually led to the other species.

For example, we have a common ancestor with chimps. This doesn't mean that our great-great...great-grandfather was a chimp. It means that chimps and we have the same great-great...great-grandfather, who was neither a human nor a chimp. Instead, he was some other species completely, and a speciation event must have occurred where his species was split and evolved in different directions. One lineage became chimps, and the other linage became humans.

This is an example of a phylogenetic tree.

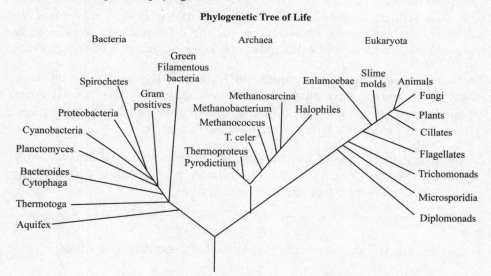

Phylogenetic Tree of Life

In this tree, the three main domains of life are seen. You can see that plants, animals, and fungi are much more closely related to each other than they are to bacteria. The archaea and the eukarya have a common ancestor with bacteria, and then they also have another common ancestor that bacteria doesn't have.

GENETIC VARIABILITY

As you know, no two individuals are identical. The differences in each person are known as **genetic variability**. All this means is that no two individuals in a population have identical sets of alleles (except, of course, identical twins). In fact, the survival of a species is dependent on this genetic variation; it allows a species to survive in a changing environment. If a population were all the same, then they would have the same weaknesses and the same strengths. Natural selection occurs only if some individuals have more evolutionary fitness and can be selected. The more variations there are among a population, the more likely that a trait will exist that might be the perfect lifesaver. The most famous example is Rudolph the Red-Nosed Reindeer. His red nose was a random mutation, but under the right conditions, it became the best nose to have. A more diverse population is more likely to survive and evolve when things are constantly changing around them.

How did all this wonderful variation come about? Well, the simple answer is that random mutations are occurring all the time. This can be because of errors by the DNA polymerase, changes to the DNA caused by transposons, or other types of DNA damage. Either way, mutations create new variations and alleles.

However, just the mutations occurring does not account for the great amount of genetic variation among a population. The mixing of genes through sexual reproduction contributes to genetic variation. During meiosis, crossing-over mixes the alleles among homologous chromosomes, and the independent assortment when chromosomes are packaged further adds to the genetic uniqueness of each gamete.

In bacteria, a process called **conjugation** increases the genetic variation, even though they are asexual reproducers. Additionally, viruses can pass around chunks of the bacterial genome during infection. This process is called **transduction**. Each of these processes will be covered more in Chapter 11.

It might be hard to think of it in this way, but genetic variation is the very foundation of evolution, as we'll soon see. Now that we've reintroduced genes, we can refine our definition of evolution. More specifically:

> Evolution is the change in the gene pool of a population over time.

The Peppered Moths

Let's take an example. During the 1850s in England, there was a large population of **peppered moths**. In most areas, exactly half of them were dark or carried alleles for dark coloring. The other half were light and carried the alleles for light coloring. This 1:1 ratio of phenotypes was observed until air pollution, due primarily to the burning of coal, changed the environment. What happened?

Imagine two different cities: City 1 (in the south) and City 2 (in the north). Prior to the Industrial Revolution, both of these cities had unpolluted environments. In both of these environments, dark moths and light moths lived comfortably side by side. For simplicity's sake, let's say our proportions were a perfect fifty-fifty, half dark and half light. However, at the height of the Industrial Revolution, City 2, our northern city, was heavily polluted, whereas City 1, our southern city, was unchanged. In the north, where all the trees and buildings were covered with soot, the light moths didn't stand a chance. They were impossible for a predator to miss! As a result, the predators gobbled up light-colored moths just as fast as they could reproduce, sometimes even before they reached an age in which they *could* reproduce. However, the dark moths with their dark alleles were just fine. With all the soot around, the predators couldn't even see them; the dark moths continued doing their thing—above all, *reproducing*. And when they reproduced, they had more and more offspring carrying the dark allele.

After a few generations, the peppered moth gene pool in City 2 changed. Although our original moth gene pool was 50 percent light and 50 percent dark, excessive predation changed the population's genetic makeup. By about 1950, the gene pool reached 90 percent dark alleles and only 10 percent light alleles. This occurred because the light moths didn't stand a chance in an environment where it was so easy to spot. The dark moths, on the other hand, multiplied just as fast as they could.

In the southern city, you'll remember, there was very little pollution. What happened there? Things remained pretty much the same. The gene pool was unchanged, and the population continued to have roughly equal proportions of light moths and dark moths.

CAUSES OF EVOLUTION

Natural selection, the evolutionary mechanism that "selects" which members of a population are best suited to survive and which are not, works both "internally" and "externally": *internally* through random mutations and *externally* through environmental pressures.

> Natural selection requires genetic variation and an **environmental pressure** that gives some individuals an advantage.

To see how this process unfolds in nature, let's return to the moth case. Why did the dark moths in the north survive? Because they were dark-colored. But how did they become dark-colored? The answer is, through **random mutation**. One day, long before the coal burning, a moth was born with dark-colored wings. As long as a mutation does not kill an organism before it reproduces (most mutations, in fact, do), it may be passed on to the next generation. Over time, this one moth had offspring. These, too, were dark. The dark- and the light-colored moths lived happily side by side until something from the outside—in our example, the environment—changed all that.

The initial variation came about by chance. This variation gave the dark moths an edge. However, that advantage did not become apparent until something made it apparent. In our case, that something was the intensive pollution from coal burning. The abundance of soot made it easier for predators to spot the light-colored moths, thus effectively removing them from the population. Therefore, dark color is an **adaptation**, a variation favored by natural selection.

Survival of the fittest is the name of the game, and any trait that causes an individual to reproduce better gives that individual **evolutionary fitness**.

Fittest of the Fit
Don't confuse evolutionary fitness with physical fitness. Often, things like strength and speed will increase evolutionary fitness, but anything that increases survival and reproducing will add to evolutionary fitness, even if it sometimes those individuals are slow or weak.

These traits are often things that simply help something to survive. After all, survival is essential for reproduction to occur. Strength, speed, height, camouflage, and many other things can be helpful. Sometimes odd things can be helpful as well. For example, a peacock's tail has been selected for because females choose to mate with those males. The tail doesn't help them to survive because it actually makes it easier for predators to catch them, but it is essential for the females to find them attractive. This would be an example of **sexual selection**.

Eventually, over long stretches of time, these two different populations might change so much that they could no longer reproduce together. At that point, we would have two different species, and we could say, definitively, that the moths had evolved. Evolution occured as a consequence of random mutation and the pressure put on the population by an environmental change. Catastrophic events can speed up natural selection and adaptation. When the rules change, sometimes the odd variations are selected for. Earth has undergone several mass extinction events. The tree of life would look much different if these events had not occurred.

Genetic drift is something that causes a change in the genetics of a population, but it is not natural selection. Instead, it is caused by random events that drastically reduce the number of individuals in a population. Often called the bottleneck effect or the founder effect, this occurs when only a few individuals are left to mate and regrow a population. If those individuals were the only survivors due to random luck, that means that the alleles that are present to regrow the population are just random alleles. They are not necessarily the fittest alleles. For example, if a hurricane swept through and only 4 birds were left on an island, the traits that they have will be the traits of the population going forward, even if those traits were not previously being selected for.

Types of Selection

The situation with our moths is an example of **directional selection**. One of the phenotypes was favored at one of the extremes of the normal distribution.

In other words, directional selection "weeds out" one of the phenotypes. In our case, dark moths were favored, and light moths were practically eliminated. Here's one more thing to remember: directional selection can happen only if the appropriate allele—the one that is favored under the new circumstances—is already present in the population. Two other types of selection are **stabilizing selection** and **disruptive selection**.

Stabilizing selection means that organisms in a population with extreme traits are eliminated. This type of selection favors organisms with average or medium traits. It "weeds out" the phenotypes that are less adaptive to the environment. A good example is birth weight in human babies. An abnormally small baby has a higher chance of having birth defects; conversely an abnormally large baby will have a challenge in terms of a safe birth delivery.

Disruptive selection, on the other hand, does the reverse. It favors both the extremes and selects *against* common traits. For example, females are "selected" to be small and males are "selected" to be large in elephant seals. You'll rarely find a female or male of intermediate size.

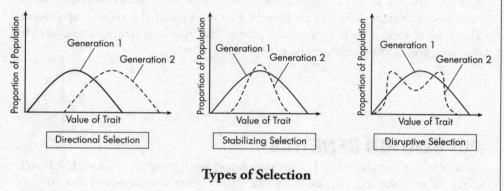

Types of Selection

SPECIES

A dog and a bumblebee obviously cannot produce offspring. Therefore, they are different **species**. However, a chihuahua and a Great Dane can reproduce (at least in theory). We would not say that they are different species; they are merely different breeds. In order for these to become different species of dogs, they would have to become **reproductively isolated** from each other. This would then allow the two groups to undergo natural selection and evolve differently. With different variation and different environmental pressures, they could each change in different ways and no longer be able to mate. This is called **divergent evolution**.

Species often originate from a common ancestor. More often than not, the "engine" of evolution is cataclysmic environmental change, such as pollution in the case of the moths. Geographic barriers, new stresses, disease, and dwindling resources are all factors in the process of evolution. Pre- and post-zygotic barriers also prevent organisms of two different species from mating to produce viable offspring. **Pre-zygotic barriers** prevent fertilization. Examples of this kind of barrier include temporal isolation, which occurs when two species reproduce at different times of the year. A **post-zygotic barrier** is related to the inability of the hybrid to produce offspring. For example, a horse and a donkey can mate to produce a mule, but mules are sterile and therefore cannot produce a second generation. Often plants undergo **speciation** due to **polyploidy**, which occurs when they get doubles of their chromosomes.

Convergent evolution is the process by which two unrelated and dissimilar species come to have similar (analogous) traits, often because they have been exposed to similar selective pressures. Examples of convergent evolution include aardvarks, anteaters, and pangolins. They all have strong, sharp claws and long snouts with sticky tongues to catch insects, yet they evolved from three completely different mammals.

There are two types of speciation: **allopatric speciation** and **sympatric speciation**. Allopatric speciation simply means that a population becomes separated from the rest of the species by a geographic barrier so that the two populations can't interbreed. An example would be a mountain that separates two populations of ants. In time, the two populations might evolve into different species. If, however, new species form without any geographic barrier, it is called sympatric speciation. This type of speciation is common in plants. Two species of plants may evolve in the same area without any geographic barrier.

POPULATION GENETICS

Mendel's laws can also extend to the population level. Suppose you caught a bunch of fruit flies—about 1,000. Let's say that 910 of them were red-eyed and 90 were green-eyed. If you allowed the fruit flies to mate and counted the next generation, we'd see that the ratio of red-eyed to green-eyed fruit flies would remain the same: 91 percent red-eyed and 9 percent green-eyed. That is, the allele frequency would remain constant. At first glance you may ask, how could that happen?

Hardy-Weinberg Equilibrium

The **Hardy-Weinberg law** states that even with all the shuffling of genes that goes on, the relative frequencies of genotypes in a population still prevail over time. The alleles don't get lost in the shuffle. The dominant gene doesn't become more prevalent, and the recessive gene doesn't disappear.

The Hardy-Weinberg law says that a population will be in genetic equilibrium only if it meets these five conditions: (1) a large population, (2) no mutations, (3) no immigration or emigration, (4) random mating, and (5) no natural selection.

When these five conditions are met, the gene pool in a population is pretty stable. Any departure from them results in changes in allele frequencies in a population. For example, if a small group of your fruit flies moved to a new location, the allele frequency may be altered and result in evolutionary changes.

Hardy-Weinberg Equations

Let's say that the allele for red eyes, R, is dominant over the allele for green eyes, r. Red-eyed fruit flies include homozygous dominants, RR, and heterozygous, Rr. The green-eyed fruit flies are recessive, rr. The frequency of each allele is described in the equation below. The allele must be either R or r. Let "*p*" represent the frequency of the R allele and "*q*" represent the frequency of the r allele in the population.

$$p + q = 1$$

This sum of the frequencies must add up to one. If you know the value of one of the alleles, then you'll also know the value of the other allele. This makes sense because Hardy-Weinberg works ONLY for populations with 2 alleles that have normal dominant-recessive behavior. So, if you know there are only 2 alleles possible and you know that 70% are the dominant allele, then the remaining 30% must be the recessive allele.

We can also determine the frequency of the *genotypes* in a population using the following equation.

$$p^2 + 2pq + q^2 = 1$$

In this equation, p^2 represents the homozygous dominants, $2pq$ represents the heterozygotes, and q^2 represents the homozygous recessives.

Both of these equations are listed on the AP Biology Equations and Formulas sheet. But how do you use them? Use the proportions in the population to figure out both the allele and genotype frequencies. Let's calculate the frequency of the genotype for green-eyed fruit flies. If 9 percent of the fruit flies are green-eyed, then the *genotype* frequency, q^2, is 0.09. You can now use this value to figure out the frequency of the recessive allele in the population. The allele frequency for green eyes is equal to the square root of 0.09—that's 0.3. If the recessive allele is 0.3, the dominant allele must be 0.7. That's because 0.3 + 0.7 equals 1.

Using the second equation, you can calculate the genotypes of the homozygous dominants and the heterozygotes. The frequency for the homozygous dominants, p^2, is 0.7 × 0.7, which equals 0.49. The frequency for the heterozygotes, $2pq$, is 2 × 0.3 × 0.7, which equals 0.42. If you include the frequency of the recessive genotype—0.09—the numbers once again add up to 1.

Don't Forget the Heterozygotes!
Remember that the number of individuals with the dominant phenotype includes those with the homozygous dominant genotype and those with the heterozygous genotype.

KEY TERMS

evolution

natural selection

Charles Darwin

Jean-Baptiste de Lamarck

paleontology

biogeography

flora

fauna

embryology

comparative anatomy

homologous structures

analogous structures

molecular biology

common ancestor

phylogenetic tree (cladogram)

genetic variability

conjugation

transduction

peppered moths

environmental pressure

random mutation

adaptation

evolutionary fitness

sexual selection

genetic drift

directional selection

stabilizing selection

disruptive selection

species

reproductively isolated

divergent evolution

pre-zygotic barriers

post-zygotic barriers

speciation

polyploidy

convergent evolution

allopatric speciation

sympatric speciation

Hardy-Weinberg law

Summary

○ Charles Darwin made the following key observations that led to his theory of natural selection:
  - Each species produces more offspring than can survive.
  - Offspring compete with each other for limited resources.
  - Organisms in every population vary.
  - The offspring with the most favorable traits are most likely to survive and reproduce.

○ Evidence for evolution includes:
  - fossils
  - biogeography
  - comparison of developmental embryology
  - comparative anatomy, including homologous and analogous structures
  - molecular biology (sequences of genes are conserved across many types of species)

○ Members of a species are defined by the ability to reproduce to produce fertile offspring. Evolution and speciation occur when environmental pressures favor traits that permit survival and reproduction of selected individuals in a varied population. The favored traits may evolve to cause divergent or convergent evolution.

○ Adaptations are random variations in traits that end up being selected for until they become prominent in the population. Genetic drift occurs when a population's traits change due to random events rather than natural selection.

○ Selection can be stabilizing if an average phenotype is preferred, disruptive if the extremes are preferred, and directional if one extreme is preferred.

○ The Hardy-Weinberg equations can be used to determine genetic variation within a population. These equations are as follows:
  - $p + q = 1$ Frequency of the dominant (p) and recessive (q) alleles
  - $p^2 + 2pq + q^2 = 1$ Frequency of the homozygous dominants (p^2), heterozygotes ($2pq$), and homozygous recessives (q^2)

○ The Hardy-Weinberg equations describe a population that is not evolving but is instead said to be in genetic equilibrium. Such a population will be so only if it is not violating any of the following five conditions (almost all populations are evolving due to a violation of one or more of these factors):
  - large population
  - no mutations
  - no immigration or emigration
  - random mating
  - no natural selection

Chapter 10 Drill

Answers and explanations can be found in Chapter 16.

1. The eye structures of mammals and cephalopods such as squid evolved independently to perform very similar functions and have similar structures. This evolution is an example of which of the following?

   (A) Allopatric speciation
   (B) Sympatric association
   (C) Divergent evolution
   (D) Convergent evolution

Questions 2–5 refer to the following graph and paragraph.

During the Industrial Revolution, a major change was observed in many insect species due to the mass production and deposition of ash and soot around cities and factories. One of the most famous instances was within the spotted moth population. An ecological survey was performed in which the number of spotted moths and longtail moths were counted in 8 different urban settings over a square kilometer in 1802. A repeat experiment was performed 100 years later in 1902. The results of the experiment are shown below.

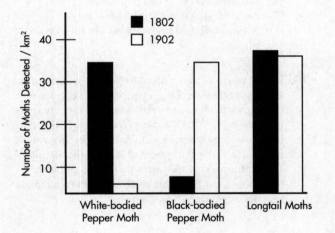

2. What type of selection is represented by the results of this study?

   (A) Stabilizing selection
   (B) Directional selection
   (C) Disruptive selection
   (D) Divergent selection

3. Which of the following statements best explains the data?

   (A) As time passed from 1802 to 1902, the frequency of white-bodied pepper moths increased and black-bodied pepper moths decreased.
   (B) As time passed from 1802 to 1902, the frequency of white-bodied pepper moths decreased and black-bodied pepper moths increased.
   (C) As time passed from 1802 to 1902, the frequency of white-bodied pepper moths and black-bodied pepper moths both increased.
   (D) As time passed from 1802 to 1902, the frequency of white-bodied pepper moths and black-bodied pepper moths both decreased.

4. Why was the population of longtail moths also surveyed in this study?

   (A) Variations in the environment were expected to alter the population of longtail moths.
   (B) Longtail moths were included as a control because they were not expected to change appreciably due to changes associated with the Industrial Revolution.
   (C) The peppered moth did not have a long enough tail to visualize.
   (D) Longtail moths were poisoned by the coal dust and suffered drastic population loss.

5. How would the results of this study have been different if factories produced white or light gray ash and soot rather than black?

   (A) There would be no change to the results of the experiment.
   (B) There would have been added selection pressure for more white-bodied spotted moths and against black-bodied spotted moths.
   (C) There would have been added selection pressure for more black-bodied spotted moths and against white-bodied spotted moths.
   (D) There would have been an increase in the frequency of both black-bodied and white-bodied spotted moths.

Question 6 represents a question requiring a numeric answer.

6. A recessive allele of a gene has a calculated frequency of 0.3 in a population. Assuming the population is in Hardy-Weinberg equilibrium, what percentage of the population is expected to be heterozygous for the gene? Give your answer to the nearest hundredths place.

7. The Middle East blind mole rat (*Nannosplalax ehrenbergi*) lives in the Upper Galilee Mountains of Israel. Two groups of these mole rats live in the same region, but scientists discovered that there is a 40% difference in mitochondrial DNA between these two groups. These rats do not seem to interbreed in the wild. This is an example of

(A) allopatric speciation
(B) sympatric speciation
(C) prezygotic barrier
(D) hybrid zone

Questions 8 and 9 refer to the following scenario.

A cruise ship with 500 people on board crashes on a deserted island. They like the island so much that they decide to stay. There is a small number of people on board with red hair, an autosomal recessive trait. Assume that red hair provides no increase in fitness.

8. What is most likely to occur with respect to the number of red-haired individuals on the island?

(A) There will be no change in the overall number of red-haired people in the population over time.
(B) Convergent evolution will occur, with all people eventually becoming red-haired.
(C) It is likely that over many generations, the proportion of red-haired people in the population will increase.
(D) Red-haired people will probably decrease in the population over many generations.

9. Evolution does not occur when a population is said to be in equilibrium by the Hardy-Weinberg definition. In the situation above, which of the following does not support the claim that evolution will occur in this population?

(A) The situation describes a small population.
(B) Mutations occur in this population.
(C) People do not usually mate randomly.
(D) Natural selection pressures do not exist on this island.

10. Peacock males have evolved to have huge, beautiful tails with numerous large eye spots. Long, colorful tails are difficult to carry when you are running from a predator. However, peahens (female peacocks) have been shown by researchers to mate preferably with males with more eye spots and longer tails, despite these traits making the males more susceptible to predation. Over time, this preference has resulted in peacocks with huge tails due to

(A) sexual selection
(B) disrupted selection
(C) divergent evolution
(D) sympatric speciation

REFLECT

Respond to the following questions:

- Which topics from this chapter do you feel you have mastered?

- Which content topics from this chapter do you feel you need to study more before you can answer multiple-choice questions correctly?

- Which content topics from this chapter do you feel you need to study more before you can effectively compose a free response?

- Was there any content that you need to ask your teacher or another person about?

Chapter 11
Animal Structure
and Function

THE STRUCTURE AND FUNCTION OF ORGANISMS

To carry on with the business of life, all higher organisms must contend with the same basic challenges: obtaining nutrients, distributing them throughout their bodies, voiding wastes, responding to their environments, and reproducing. To accomplish these basic tasks, nature has come up with solutions.

Unicellular and multicellular organisms alike meet these challenges through organization and specialization. We already learned about the organelles within a cell allowing compartmentalization. Each organelle does it own task, and they work together to keep the cell alive. In many animals and humans, there are higher levels of organization.

Homeostasis

All living things have a "steady state" that they try to maintain. This is helpful because the structures and processes within living things are sensitive to things like temperature, pH, pressure, salinity, osmotic pressure, and many other things. The set of conditions that living things can live in successfully is called **homeostasis**. The body is constantly working to maintain this state by taking measurements and then responding appropriately. That might mean shivering if it is cold or releasing the hormone insulin if blood glucose is high. Depending on the condition, the body will respond accordingly.

Many of these responses are controlled by negative or positive feedback pathways. A **negative feedback pathway** (also called **feedback inhibition**) works by turning itself off using the end product of the pathway. The end product inhibits the process beginning again, thus shutting down the pathway. This is a common strategy to conserve energy. If plenty of X is made, then the pathway to make X can be turned off. Sometimes, a very distant end product will turn off a process occurring several steps back. For example, ATP turns off the beginning stages of glycolysis, although ATP is not made until much later in cellular respiration.

A **positive feedback pathway** also involves an end product playing a role, but instead of inhibiting the pathway, it further stimulates it. For example, once X is made, it tells the pathway to make even more X. This is less common, but it occurs during fruit ripening and labor and delivery, which ramps up and up as it proceeds.

DEVELOPMENT

For the AP test, it is important to have a general idea of how the body works. Let's start at the beginning.

Embryonic Development

How does a tiny, single-celled egg develop into a complex, multicellular organism? By dividing, of course. The cell changes shape and organization many times by going through a succession of stages. This process is called **morphogenesis**.

When an egg is fertilized by a sperm, it forms a diploid cell called a **zygote.**

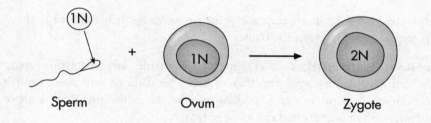

Sperm   Ovum   Zygote

Fertilization triggers the zygote to go through a series of divisions. As these occur, the embryo becomes increasingly differentiated, or specialized. An undifferentiated cell is like a blank slate—it can become any type of cell. Once a cell begins to specialize, it is limited in its future options.

In order for a cell to differentiate, it must change. Certain genes will be expressed, and other genes might be turned off. Cells called organizers release signals that let each cell know how they should develop. As an over-simplified example, a cell destined to become a muscle might have an increase in things that make it flexible, while something destined to become bone might have a decrease in those things. Once the change has been made, the future muscle cell can't change its mind and become a bone cell.

The early genes that turn certain cells in the early embryo into future-this or future-that are called **homeotic genes**. A subset of homeotic genes are called **Hox genes**. The timing is essential for the induction (or activation) of these genes. Exactly the right bit of the embryo must be modified at exactly the right time, or the embryo may develop with the brain in the wrong place or too many limbs or only limbs on one side of the body. Severely damaged embryos will stop developing.

An interesting additional tool that the developing embryo uses is apoptosis, or programmed cell death. Certain bits of the developing embryo are used as a scaffold for development (like the cells that used to exist in the spaces between your fingers and toes), and when they are no longer necessary, those cells commit suicide. Think of it like an eraser coming along and erasing the webbing tissue between fingers and toes.

BODY SYSTEMS

Once the embryo develops into a fully formed organism, it is highly specialized. Groups of cells that all perform the same function are called a **tissue,** and several tissues can come together to form specialized structures called **organs**. Several organs then come together to form **body systems**.

Fortunately, you need to know details about only the **immune system,** the **nervous system,** and the **endocrine system.** Here is a brief list of other major systems so you get the gist of their functions:

Circulatory: Blood vessels (arteries, capillaries, and veins) carry blood around the body to transport chemical signals and to bring supplies to cells and carry waste away. The blood flow is controlled by the heart, which pumps the blood through the blood vessels.

Respiratory: The lungs are responsible for gas exchange (O_2 and CO_2). They also help maintain pH levels in the blood.

Digestive: The esophagus, stomach, small intestine, large intestine, pancreas, liver, and gall bladder work together to break food down and absorb nutrients. The stomach does mixing and breakdown. The small intestine does most of the absorption. The pancreas makes lots of enzymes.

Excretory: The kidneys filter the blood and reabsorb things that the body wants to keep and get rid of the rest in urine.

Reproductive: The male and female systems are largely controlled by hormones and allow the production of gametes and the ability to reproduce.

Muscular and Skeletal: The body has skeletal muscle and smooth muscles that each contract by action potential signals from the nervous system. The skeletal system provides structure, protection, and calcium storage.

THE IMMUNE SYSTEM

The immune system is the body's defense system. It is a carefully and closely co-ordinated system of specialized cells, each of which plays a specific role in the war against bodily invaders. Let's take a brief sidetrack and talk a bit about these invaders. **Pathogens** are disease-causing biological agents that can generally be divided into bacteria, viruses, fungi, and parasites.

Bacteria

We have been talking on and off about bacteria already. They are prokaryotes that come in many shapes and sizes. They can infect many things, and sometimes they cause harm and sometimes they do not. You may have heard of "gut bacteria" before; this is a special colony of bacteria that lives inside each one of us. It helps us with some digestion and makes some things we need. We have a mutualistic relationship with our gut bacteria.

Bacteria divide by fission; however, this does not increase their genetic diversity. Instead, they can perform **conjugation** with other bacterial cells and swap some of their DNA. Genetic variety among bacteria is leading to increased antibiotic resistance. As we mentioned in Chapter 10, an increase in genetic variation increases the likelihood that a population will survive a catastrophic event.

Viruses

Viruses are nonliving agents capable of infecting cells. Why are viruses considered nonliving? They require a host cell's machinery in order to replicate. A virus consists of two main components: a protein capsid and genetic material made of DNA or RNA, depending on the virus. Viruses are all very specific in which type of cells they infect, and the thing infected by a virus is called a **host.**

Viruses have one goal: replicate/spread. In order to do this, a virus needs to make more genome and make more capsid. They then assemble together into new viral particles.

The viral genome carries the genes for building the capsid and anything else the virus needs that the host cannot provide. Sometimes, if two viruses infect the same cell, there will be mixing of the genomes, especially if the viruses that have genomes split between several chromosome-like segments. A new virus particle might emerge that is a blend of the two viruses.

A commonly studied virus is a **bacteriophage** (a virus that infects bacteria). Bacteriophages undergo two different types of replication cycles, the **lytic cycle** and the **lysogenic cycle**. In the lytic cycle, the virus immediately starts using the host cell's machinery to replicate the genetic material and create more protein capsids. These spontaneously assemble into mature viruses and cause the cell to lyse, or break open, releasing new viruses into the environment. In the lysogenic cycle, the virus incorporates itself into the host genome and remains dormant until it is triggered to switch into the lytic cycle. A virus can hide in the genome of a bacteria cell for a very long time. During this time, the cell may divide and replicate the virus as well. By the time the lytic cycle is triggered, the virus may have been replicated many many times as the cell hosting it divides.

When a virus excises from a host genome (becomes unintegrated), it sometimes accidentally takes some of the bacterial cell's DNA with it. Then, that gets replicated and packaged into new viral particles with the viral genome. The next cell that gets infected is not only getting infected with the viral genome, but also with that stolen chunk of bacterial DNA. If that chunk held a gene for something like antibiotic resistance, the next cell that gets infected will gain that trait. The transfer of DNA from a virus to a bacterial cell is called transduction.

Viruses that infect animals do not have to break their way out of the cell the same way that a bacteriophage does. Since animals cells don't have a cell wall, the viruses often just "bud" out of the membrane like exocytosis. When a virus does this, it becomes enveloped by a chunk of cell membrane that it takes with it. Viruses with a lipid envelope are called **enveloped viruses**.

Retroviruses like HIV are RNA viruses that use an enzyme called **reverse transcriptase** to convert their RNA genomes into DNA so that they can be inserted into a host genome. RNA viruses have extremely high rates of mutation because they lack proofreading mechanisms when they replicate their genomes. This high rate of mutation will create lots of variety, which makes these viruses difficult to treat. As was discussed in Chapter 10, they evolve quickly as drug-resistant mutations become naturally selected. New drugs must constantly be identified to treat the resistance.

The Two Immune Responses

Foreign molecules—be they viral, bacterial, or chemical—that can trigger an immune response are called **antigens**. Humans and other vertebrates have two types of immune responses: the **innate immune response** and the **adaptive/specific immune response**. The innate is more of a general anti-invader response. The body recognizes certain things that are common of foreign things and destroys them. The adaptive response carefully catalogs and handles each antigen in a particular way. It also has a memory component that helps it fight repeat attackers very efficiently.

The Innate Immune System

The body's first line of defense against foreign substances is the skin and the mucous lining of the respiratory and digestive tracts. If these defenses are not sufficient, other nonspecific defense mechanisms are activated. These include **phagocytes** such as **macrophages** (which engulf antigens), **complement proteins** (which lyse the cell wall of the antigen), **interferons** (which inhibit viral replication and activate surrounding cells that have antiviral actions), and **inflammatory responses** (a series of events in response to antigen invasion or physical injury). Activation of these defenses requires immune cells to recognize the foreign substance when it successfully binds to the immune cell's receptor and activate intracellular signaling pathways. This is another example of how important cell communication is to survival. These first line defenses are called the innate immune system, and many organisms have this level of defense. The specifics vary with organism structure, but there are several common themes. For example, a thick outer surface usually protects the organism, such as the skin of a human, the cell wall of a bacteria, or the cuticle of a plant. In all organisms, cell communication pathways are essential to recognize "self" cells, recognize foreign substances, and initiate immune responses when under attack by a pathogen (a disease-causing agent) or chemical.

The Adaptive/Specific Immune Response

Lymphocytes are the primary cells of the immune system. They are found mostly in the blood and lymph nodes and are a type of white blood cell (or leukocyte). There are two types of lymphocytes: B-cells and T-cells. When an individual becomes infected by a pathogen, B- and T-lymphocytes get activated.

B-lymphocytes mature in bone marrow and are involved in the **humoral response**, which defends the body against pathogens present in extracellular fluids, (or humors) such as lymphatic fluid or blood. Each B-cell has a special receptor on its surface that can bind only to foreign antigens. If a pathogen arrives that fits the receptor, the B-cell becomes activated (with the help of a T-cell). The B-cell will begin to replicate and seek out more of that pathogen. Some B-cells become **memory B-cells** that remain in circulation, allowing the body to mount a quicker response if a second exposure to the same pathogen should occur. Other B-cells become **plasma cells** that produce **antibodies**, which are specific proteins that bind to the same antigen that originally activated the B-cell. The antibody that is made by each B-cell is identical to the the surface receptor that originally caught the antigen. Each B-cell has a unique receptor/antibody that it makes.

All antibodies have the same basic monomer structure that is shaped like the letter Y. The stem of the Y is always the same and can interact with other cells in the immune response. The arms of the Y are always unique because this is where the antibody binds antigen. On each antibody, both arms bind the same shape so that it can hold two antigens at once. As mentioned, this Y shape is just the monomer of an antibody. Sometimes antibodies can have one, two, or five Y-monomers with the stem parts attaching and the arms facing outward. Antibodies are typically made only when the appearance of antigens in the body stimulates a defense mechanism, but they will linger around after an infection. When an antibody binds to an antigen, it marks the antigen for destruction. Antibodies can also combat the antigen just by binding to it because this alone can wreak havoc for the pathogen.

Antigen Binding Site

Variable Region

Constant Region

T-lymphocytes, maturing in the thymus, are involved in **cell-mediated immunity**; this branch of the immune system is responsible for monitoring "self" cells to make sure they are still healthy. The plasma membrane of cells has **major histocompatibility complex (MHC) markers** that allow T-cells to get a glimpse of what is happening inside each cell. T-cells have special antigen-recognizing receptors just like B-cells, and if the receptors see something foreign, the T-cells jump into action.

MHC I are on all nucleated cells. They take peptides that have been found inside the cell and hold them up on the surface so that a type of circulating T-cell called a **cytotoxic T-cell** can inspect them. If they decide the cell is infected or is a tumor cell, they signal for it to be destroyed by **apoptosis**, which is a careful self-destruct program.

MHC II are on special immune cells, like B-cells and macrophages, that identify and engulf antigens. These types of cells are called antigen-presenting cells. They hold up a chunk of whatever antigen the immune cell has picked up and they let a type of T-cell called a **helper T-cell** inspect it. If the T-cell agrees that it is something

Remember, B-cells look for things out in the blood and lymph and can make antibodies. T-cells check out what is inside cells. Cytotoxic T-cells check all cells, and Helper T-cells check what immune cells have picked up and help activate them.

If you have a measurable level of antibodies against something, this means you have been infected with it at some point. It doesn't necessarily mean you are currently infected.

foreign, it helps activate the immune cell. Think of the helper T-cell as a double-checker. This interaction between different branches of the immune system requires B-cells and T-cells to recognize each other and coordinate an immune attack—another process tightly controlled by cell signaling and communication pathways.

A **lymph node** is a mass of tissue found along a lymph vessel. A lymph node contains a large number of lymphocytes. Because lymphocytes are important in fighting infection, they multiply rapidly when they come in contact with an antigen. As a consequence, lymph nodes swell when they're fighting an infection. That's why when you have a sore throat, one of the first things a doctor does is touch the sides of your throat to see if your lymph nodes are swollen, a probable sign of infection.

One thing to remember about immune cells and blood cells: all blood cells, white and red, are produced in the bone marrow. However, red blood cells (or erythrocytes) play no role in the immune system; instead, they help transport oxygen throughout the body. They are full of hemoglobin.

Vaccines

After an infection, the adaptive immune response keeps around groups of cells called memory B-cells and memory T-cells. They act as a militia and keep their eyes peeled for the same antigens they have seen before. The second time that you are infected with a pathogen, you will usually fight it off very quickly. You might not even know you were even infected because your body destroys it so fast. Vaccines are tiny doses of antigen that have been modified so they are not dangerous. Your immune system is introduced to them and prepares a militia, but you don't ever get the "real" infection. If you do come into contact with the real version, your body will fight it off ASAP. In a nutshell, getting vaccinated is getting a tiny harmless dose now so you don't get knocked off your feet by the big, bad version later.

AIDS

AIDS, or acquired immunodeficiency syndrome, is a devastating disease that interferes with the body's immune system. AIDS is caused by the specific infection of helper T-cells by human immunodeficiency virus (HIV). The virus essentially wipes out helper T-cells, preventing the body from defending itself. Those afflicted with AIDS do not die of AIDS itself, but rather of infections that they can no longer fight off, due to their compromised immune systems.

Auto-Immunity

The B-cells and T-cells are carefully created so that they do not react to "self" and only react to foreign things. However, sometimes mistakes happen and the immune system begins to think that "self" is dangerous. These attacks cause **auto-immune diseases** such as Type I diabetes, rheumatoid arthritis, lupus, and celiac disease.

THE NERVOUS SYSTEM

All organisms must be able to react to changes in their environment. As a result, organisms have evolved systems that collect and process information from the outside world. The task of coordinating this information falls to the nervous system. The simplest nervous system is found in the hydra. It has a **nerve net** made up of a network of nerve cells, which sends impulses in both directions. As animals became more complex, they developed clumps of nerve cells called **ganglia**. These cells are like primitive brains. More complex organisms, like humans, have a brain with specialized cells called **neurons**.

Neurons

The functional unit in the nervous system is a neuron. Neurons receive and send the neural impulses that trigger organisms' responses to their environments. Let's talk about the parts of a neuron. A neuron consists of a **cell body**, **dendrites**, and an **axon**.

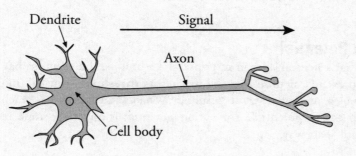

The cell body contains the nucleus and all the usual organelles found in the cytoplasm. Dendrites are short extensions of the cell body that receive stimuli. The axon is a long, slender extension that transmits an impulse from the cell body to another neuron or to an organ. A nerve impulse begins at the top of the dendrites, passes through the dendrites to the cell body, and moves down the axon.

How Neurons Communicate

Within a neuron, the signal is called an **action potential**, which is a wave of positive charge that sweeps down the axon. In order for the signal to be clear, the cell must have a "normal," non-positive state. Nearly all cells in the body have a natural resting state that is negative. The natural charge of a cell is called the **resting membrane potential**, which represents the difference in charge from inside the cell to outside the cell.

The resting potential arises from two activities:

- **The Na^+K^+-ATPase pump:** This pump pushes two potassium ions (K^+) into the cell for every three sodium ions (Na^+) it pumps out of the cell, which leads to a net loss of positive charges within the cell.
- **Leaky K^+ channels:** Some potassium channels in the plasma membrane are "leaky," allowing a slow diffusion of K^+ out of the cell.

Both the Na⁺K⁺-ATPase pump and the leaky channels cause a potential difference between the inside of the neuron and the surrounding interstitial fluid. The resting membrane potential is always negative inside the cell, and the neuronal membrane is said to be **polarized**. In humans, the negative charge is −70mV.

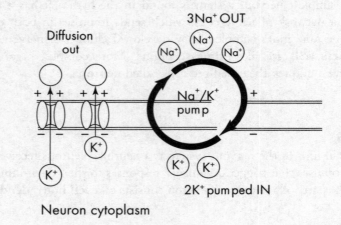

Action Potential

Here's what a neuron does in response to a stimulus. If a stimulus has enough intensity to excite a neuron, the cell reaches its **threshold**—the minimum amount of stimulus a neuron needs to respond. When the threshold is reached, the cell "fires" an action potential. The action potential is an **all-or-none response**—it doesn't fire "part way."

(1) **Reaching threshold:** An outside stimulus causes a slight influx of positive charge in the neuron. When this influx causes the cell body of the neuron to reach −50mV, threshold is reached.

(2) **Sodium channels open:** The −50mV voltage causes the opening of many voltage-gated sodium channels near where the axon meets the cell body. Remember, the sodium potassium pump is actively pumping sodium out of the cell. This means that it is pumping sodium against the concentration gradient, meaning there is already a lot of sodium outside the cell. Therefore, the sodium wants to flow into the cell. When these channels are opened at threshold, a ton of sodium rushes into the cell. This causes the inside of the cell to become very positive (since sodium is positively charged). This is called **depolarization**. When the cell reaches +35mV inside, these sodium channels close again.

(3) Potassium channels open: As the sodium channels close, the potassium channels open. Potassium flows the opposite direction of sodium, out of the cell. This outflux of positives causes the cell to become negative again. This is called **repolarization**. When the voltage reaches around –90mV, the potassium channels close, and the cell returns to resting membrane potential.

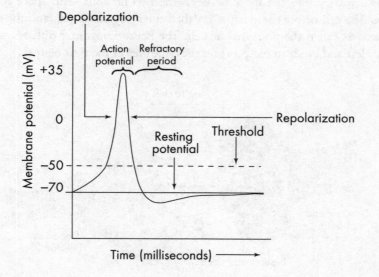

The Refractory Period

The period after an action potential is known as the **refractory period**. There are two different refractory periods. The first is caused because the sodium channels are unable to open again right away. They are in lockdown mode for a short time. The second is caused because the cell dips down below the resting membrane potential and is super negative. A greater stimulus is required to reach the threshold, so it is more difficult to initiate another action potential.

To summarize:

- A "resting" neuron is polarized—that is, the charge is more negative on the inside of the cell than on the outside.
- When an action potential comes along, the neuron transmits an impulse down its axon.
- First, voltage-gated sodium channels open, allowing sodium ions to rush in. This is known as depolarization: the neuron becomes more positive on the inside and more negative on the outside.
- Sodium channels close, and potassium channels open, which restores its negative charge. This is known as repolarization.
- The neuron enters a refractory period.
- The neuron reestablishes the ion distribution thanks to the sodium-potassium pump.

Passing Along the Signal

When one small area is depolarized, it causes a "domino effect." The action potential spreads to the rest of the axon. The impulse is transmitted down the axonal membrane until it reaches the end of the axon called the **axon bulb**. Let's discuss how the neuron manages to pass the impulse to the next neuron.

The Other Synapse
This page explains a chemical synapse, which is the most talked about type of synapse. However, a gap junction between cells is also a type of synapse called an electrical synapse. The cells in the heart are connected by this type of synapse.

When an impulse reaches the end of an axon, the axon releases a chemical called a **neurotransmitter** into the space between the two neurons. This space is called a **synapse.** The cell before the synapse is called the presynaptic cell, and the cell after the synapse is called the postsynaptic cell. The neurotransmitter diffuses across the synaptic cleft and binds to receptors on the dendrites of the next neuron.

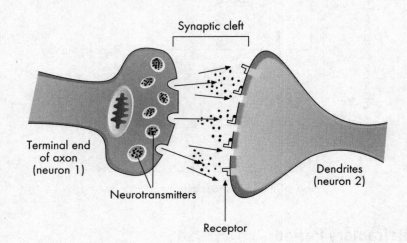

This will trigger an action potential in the second neuron if the postsynaptic membrane is excited. However, sometimes the signal will cause the postsynaptic membrane to become negatively charged (inhibited). Now the impulse moves along the second neuron from dendrites to axon.

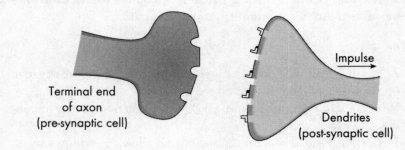

There are many neurotransmitters, but the most important one for the AP Biology Exam is called **acetylcholine**. Acetylcholine is a neurotransmitter that:

- is released from the end of an axon when Ca^{2+} moves into the terminal end of the axon
- is picked up almost instantly by the dendrites of the next neuron
- can stimulate muscles to contract
- is released between neurons in the parasympathetic system, which we will discuss shortly

The extra acetylcholine in the synaptic cleft is broken down by the enzyme **acetylcholinesterase**. Other important neurotransmitters include **norepinephrine** and **GABA**.

Speed of an Impulse

Sometimes a neuron has supporting cells that wrap around its axon. These cells are called **Schwann cells**. Schwann cells produce a substance called the **myelin sheath**, which insulates the axon.

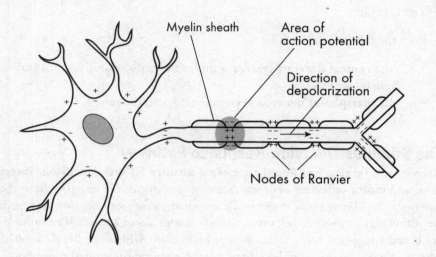

As you can see from the illustration above, the whole axon isn't covered with myelin sheaths. The spaces between myelin sheaths—the exposed regions of the axon—are called the **nodes of Ranvier**. Myelin sheaths speed up the propagation of an impulse. Instead of the standard "domino effect" that occurs during an action potential, the impulse can now jump from node to node. It is almost as if there are giant dominoes which can skip sections of the neuron. This form of conduction is called **saltatory conduction**.

Thanks to the myelin sheath, the neuron can transmit an impulse down the axon far more rapidly than it could without its help.

Parts of the Nervous System

Central Nervous System

The nervous system can be divided into two parts: the central nervous system and the peripheral nervous system.

All of the neurons within the brain and spinal cord make up the central nervous system. All of the other neurons lying outside the brain and the spinal cord—in our skin, our organs, and our blood vessels—are collectively part of the peripheral nervous system. Although both of these systems are really part of one system, we still use the terms *central* and *peripheral*. The brain has many regions and parts that each have specific functions, but you do not need to know the details of the different sections.

So keep the following in mind:

- The **central nervous system** includes the neurons in the brain and spinal cord.
- The **peripheral nervous system** includes all the rest.

The Stimulus-Decision-Response Pathway

Neurons can be classified into three groups: **sensory (affector) neurons**, **interneurons**, and **motor (effector) neurons**. Sensory neurons receive impulses from the environment and bring them to the body. For example, sensory neurons in your hand are stimulated by touch. They transmit the message about the touch stimulus to the brain and the spinal cord. Once there, interneurons will make the decision about what to do about the stimulus. Then, motor neurons transmit the decision from the brain to muscles or glands to produce a response. The muscle will respond by contracting, or the gland will respond by secreting a substance (such as a hormone).

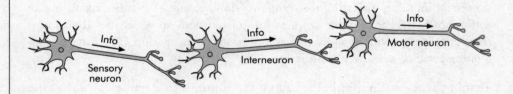

So, the basic function of the nervous system is to detect a stimulus and transmit it to the central nervous system, which makes a decision and then sends that decision out to the body to respond to the stimulus.

Reflexes

As mentioned, the central nervous system is composed of the brain and the spinal cord. Usually, the brain is the part that makes the decisions, so most stimuli are sent to the brain. However, in certain instances, the spinal cord is capable of making the decision. This is what happens in **reflexes.** A quick decision is required to prevent injury, and reflex arcs are pre-determined decisions that the spinal cord is set up to execute. The spinal cord may make a decision to withdraw a hand because of pain or to relax a muscle when its opposite contracts to prevent injury.

THE ENDOCRINE SYSTEM

Like the nervous system, the endocrine system is an excellent example of signaling. It is responsible for maintaining homeostasis and for coordinating responses to many stimuli in the body. Chemical messengers can be produced in one region of the body to act on target cells in another region. These chemicals, known as **hormones**, are produced in specialized organs called **endocrine glands**. Hormones have a number of functions including regulating growth, behavior, development, and reproduction. Other chemical messengers are used for communication. For example, **pheromones** help animals to communicate with members of their species and attract the opposite sex.

Endocrine glands release hormones directly into the bloodstream, which carries them throughout the body. Take a look at the endocrine glands in the human body.

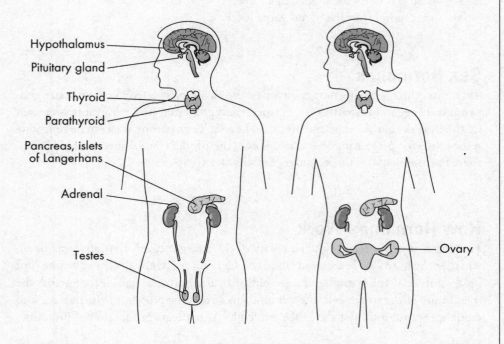

Although hormones flow in your blood, they affect only specific cells. The cells that a hormone affects are known as the **target cells**. Suppose, for example, that gland X makes hormone Y. Hormone Y, in turn, has some effect on organ Z. We would then say that organ Z is the *target organ* of hormone Y.

Hormones also operate by a negative feedback system. That is, an excess of the hormone will signal the endocrine gland to temporarily shut down production. For example, when hormone Y reaches a peak level in the bloodstream, the organ secreting the hormone, gland X, will get a signal to stop producing hormone Y. Once the levels of hormone Y decline, the gland can resume production of the hormone. Now we will go through a few examples of hormone regulation.

The Pancreas

The pancreas secretes two well-known hormones, **glucagon** and **insulin**, both of which are produced in clusters of cells called the **islets of Langerhans**. The target organs for these hormones are the liver and muscle cells. Glucagon, produced by α cells, stimulates the liver to convert **glycogen** into glucose and to release that glucose into the blood. Glucagon therefore *increases* the levels of glucose in the blood. Insulin has precisely the opposite effect that glucagon does.

When the blood has too much glucose floating around, insulin, produced by β cells, allows body cells to remove glucose from the blood. Consequently, insulin *decreases* the level of glucose in the blood. Insulin is particularly effective on muscle and liver cells. In short:

- insulin lowers the blood sugar level
- glucagon raises the blood sugar level

Sex Hormones

There are three key hormones involved in the reproductive system: **estrogen**, **progesterone**, and **testosterone**. Estrogen and progesterone are hormones released by the ovaries that regulate the menstrual cycle. Testosterone is the male hormone responsible for promoting spermatogenesis, the production of sperm. In addition, these hormones maintain secondary sex characteristics.

How Hormones Work

How do hormones trigger the activities of their target cells? That all depends on whether the hormone is a steroid (lipid soluble) or a protein, peptide, or amine (not lipid soluble). If the hormone is a steroid, then the hormone can diffuse across the membrane of the target cell. It then binds to a receptor protein in the nucleus and regulates transcription of the DNA, which in turn affects production of proteins.

However, if the hormone is a protein, peptide, or amine, it can't get into the target cell by means of simple diffusion. Remember: only hydrophobic things can pass through the membrane. The hormone binds to a receptor protein on the *cell membrane* of the target cell. This often causes a signal cascade whereby the signal is transduced into the cell without the actual hormone itself ever entering the cell.

Here's a summary of the hormones and their effects on the body.

ORGAN	HORMONES	EFFECT
Anterior Pituitary	FSH	Stimulates activity in ovaries and testes
	LH	Stimulates activity in ovary (release of ovum) and production of testosterone
	ACTH	Stimulates the adrenal cortex
	Growth Hormone	Stimulates bone and muscle growth
	TSH	Stimulates the thyroid to secrete thyroxine
	Prolactin	Causes milk secretion
Posterior Pituitary	Oxytocin	Causes uterus to contract
	Vasopressin	Causes kidney to reabsorb water
Thyroid	Thyroid Hormone	Regulates metabolic rate
	Calcitonin	Lowers blood calcium levels
Parathyroid	Parathyroid Hormone	Increases blood calcium concentration
Adrenal Cortex	Aldosterone	Increases Na^+ and H_2O reabsorption in kidneys
Adrenal Medulla	Epinephrine Norepinephrine	Increase blood glucose level and heart rate
Pancreas	Insulin	Decreases blood sugar concentration
	Glucagon	Increases blood sugar concentration
Ovaries	Estrogen	Promotes female secondary sex characteristics and thickens endometrial lining
	Progesterone	Maintains endometrial lining
Testes	Testosterone	Promotes male secondary sex characteristic and spermatogenesis

You are not required to know the information on this table, but it might come in handy as an example to use in a free-response question.

While the nervous system and the endocrine system work in close coordination, there are significant differences between the two:

- The nervous system sends nerve impulses using neurons, whereas the endocrine system secretes hormones.
- Nerve impulses control rapidly changing activities, such as muscle contractions, whereas hormones deal with long-term adjustments.

KEY TERMS

homeostasis
negative feedback pathway (feedback inhibition)
positive feedback pathway
embryonic development
morphogenesis
zygote
fertilization
homeotic genes/Hox genes
tissue
organ
body systems

Immune System

pathogen
bacteria
conjugation
viruses
host
bacteriophage
lytic cycle
lysogenic cycle
enveloped virus
retrovirus
reverse transcriptase
antigen
innate immune system
adaptive and specific immune response
phagocytes
macrophages
complement proteins
interferons
inflammatory response
lymphocytes
B-lymphocytes
humoral response
memory B-cell
plasma cell
antibody
T-lymphocytes
cell-mediated immunity
major histocompatibility complex
   (MHC) markers
cytotoxic T-cell
apoptosis
helper T-cell
lymph node
AIDS
auto-immune disease

Nervous System

nerve net
ganglia
neurons
cell body
dendrites
axon
action potential
resting membrane potential
Na^+/K^+ ATPase pump
leaky K^+ channels
polarized
threshold
all-or-none response
depolarization
repolarization
refractory period
axon bulb
neurotransmitter
synapse
acetylcholine
aceytlcholinesterase
norepinephrine
GABA
Schwann cells
myelin sheath
nodes of Ranvier
saltatory conduction
central nervous system
peripheral nervous system
sensory (affector) neurons
interneurons
motor (effector) neurons
reflexes

Endocrine System

hormones
endocrine glands
pheromones
target cells
glucagon (produced in alpha cells)
insulin (produced in beta cells)
islets of Langerhans
glycogen
estrogen
progesterone
testosterone

Summary

- The AP Biology Exam requires specific knowledge of only three systems: immune system, nervous system, and endocrine system. Other body systems might appear on the test, but only in the context of the kinds of themes covered in this chapter: structure and function relationships, communication between cells, and so on. Understanding the following key points about the three systems highlighted in this chapter will help you with the thematic questions about various body systems on the test.

Immune System

- The immune system is the defense system against pathogens that enter the organism. Bacteria and viruses are common pathogens. Bacteriophages are viruses that infect bacteria. Viruses require a host to replicate and sometimes lyse the host cell during infection. There are two sides to the immune system:
  - The innate immune system is the first response, the initial protection, and it includes the skin and mucous linings. Phagocytes and complement proteins also target this initial response to neutralize invaders.
  - Acquired immunity can be cell mediated or antibody based. T-lymphocytes include cytotoxic killer T-cells and helper T-cells. B-cells include plasma and memory B-cells.

Nervous System

- Neurons are highly specialized, terminally differentiated cells. They are composed of dendrites, cell bodies, axons, and synaptic terminals:
  - Neurons are classified into three groups: sensory neurons, motor neurons, and interneurons.
  - Neuronal transmission is based on action potentials, which are temporary disruptions of the ion concentrations set up by the resting membrane potential.
  - The junction between neurons is called a synapse. The presynaptic axon releases neurotransmitters to the postsynaptic dendrite to perpetuate the transmission.
  - Myelin speeds the transmission along axons.

- The central nervous system is composed of the brain and spinal cord.

- The peripheral nervous system senses and responds to stimuli.

Endocrine System

- Endocrine glands are tissues or organs that excrete hormones. Hormones mediate growth, reproduction, waste disposal, nutrient absorption, and behavior.

- Target cells receive the specific hormone via receptors either on the surface of the cell or internally and react through signal transduction and response.

Chapter 11 Drill

Answers and explanations can be found in Chapter 16.

1. Lymph nodes are often evaluated when a patient is believed to have an active infection or cancer. Why?

   (A) Active infections or cancer often generate large quantities of fluids as cells rupture; the lymph nodes often swell as a result of the buildup of these fluids.
   (B) During an active infection, lymphocytes expand rapidly in lymph nodes, causing swelling of the structures.
   (C) Lymph nodes are the most common tissue site of active infections or cancer.
   (D) Cancer often metastasizes to lymph nodes, causing them to expand as new cancerous cells multiply.

Questions 2-4 refer to the following chart and paragraph.

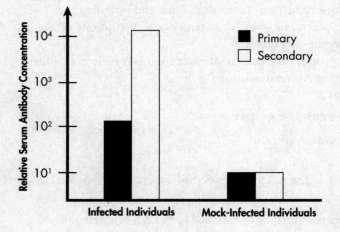

A scientist wishes to test the efficacy and immunogenicity of a newly developed, live-attenuated virus vaccine strain. To test for immunogenicity, the scientist acquires 100 healthy volunteers and exposes them to either the vaccine strain or a placebo strain (consisting of isotonic saline solution). Twelve days following the initial primary exposure, the serum antibody concentration is determined by an ELISA analysis. A follow-up exposure to test for existing immunity is performed on day 60 using the same 100 individuals. Serum antibody titers are determined for the secondary challenge on day 72. These data are shown in the chart above.

2. The placebo strain was included in the experimental design in order to

   (A) evaluate the immune response to isotonic saline solution
   (B) determine primary and secondary immune responses to the virus
   (C) compare antibody titers of infected and mock-infected individuals as a control
   (D) practice before challenging the same individuals with the vaccine strain

3. What type of immune cells are responsible for generating the antibodies in this experiment?

   (A) Erythrocytes
   (B) B-cells
   (C) T-cells
   (D) Phagocytes

4. How can the scientist increase the statistical significance of these data?

   (A) She could repeat the experiment with a greater sample size.
   (B) She could use more control variables.
   (C) She could perform a third challenge on the same 100 individuals.
   (D) She could use a different statistical test, which provides the necessary statistical significance.

5. Innate immune system defenses in animals include skin, mucous membranes, and defensive enzymes present in saliva, tears, stomach acid, and skin. Analogous structures in plants could include

   (A) long stems
   (B) thorns and spines
   (C) attractive scents
   (D) stomata

6. When lymphocytes first encounter a pathogen, receptors on their surface first bind to the outside of the pathogen. Next, this activated lymphocyte multiplies and becomes numerous, allowing them to bind to many more pathogens of the same type in the bloodstream. What are other results from this initial recognition of the pathogen?

   (A) Antibodies specific to the pathogen are secreted by B-cells.
   (B) The pathogen is destroyed by helper T-cells.
   (C) Only plasma B-cells remain.
   (D) The pathogen is permitted to infect more cells prior to its destruction.

7. The image below depicts an action potential occurring in an active neuron. Which of the following best explains why the membrane potential depolarizes at –50 mV?

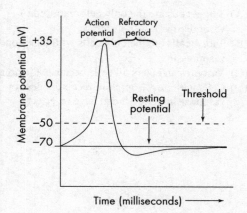

   (A) At –50 mV, voltage-gated potassium channels open, allowing potassium ions (K⁺) to rush into the cell.
   (B) At –50 mV, voltage-gated sodium channels open, allowing sodium ions (Na⁺) to rush into the cell.
   (C) At –50 mV, voltage-gated potassium channels open, allowing potassium ions (K⁺) to rush out of the cell.
   (D) At –50 mV, voltage-gated sodium channels open, allowing sodium ions (Na⁺) to rush out of the cell.

8. Fugu is a Japanese delicacy made from the preparation of pufferfish. Pufferfish produce a powerful toxin called tetrodotoxin, which binds to and blocks voltage-gated sodium channels. Specially trained Japanese chefs must carefully remove the toxin during preparation of fugu to prevent inadvertent poisoning by ingestion. Which of the following would likely occur to a neuron exposed to tetrodotoxin?

   (A) The neuron would be able to send action potentials but would not be able to release neurotransmitters to neighboring neurons.
   (B) The neuron would be able to send action potentials but would not be able to receive neurotransmitters from a presynaptic neuron.
   (C) The neuron would not be able to send action potentials and would be unable to continue a neural response.
   (D) The neuron would not be able to send action potentials but will still be depolarized by the influx of sodium ions.

9. Ouabain is historically used by Somali tribes on poisoned arrows. Ouabain inhibits the activity of the sodium-potassium pump and is therefore used as a powerful neurotoxin. What immediate effect would ouabain have on motor and sensory neurons in a victim of a poisoned arrow?

   (A) The muscles would spasm because of an excess amount of calcium released from the sarcoplasmic reticulum.
   (B) Action potentials would not be capable of firing because the pumps are blocked and the cell would be very positively charged.
   (C) Brain damage would result as the poison spread up the neurons to the central nervous system.
   (D) Action potentials would fire in the affected neurons, and then repolarization would be impossible without functioning sodium-potassium pumps. Paralysis would result.

10. In a neuron, the majority of the protein manufacturing machinery will be found in the

    (A) dendrites
    (B) cell body
    (C) axon
    (D) axon terminal

11. Acetylcholinesterase is an enzyme hydrolase that destroys the neurotransmitter acetylcholine in the neuromuscular junction. Patients with defective acetylcholinesterase will have

    (A) more easily relaxed musculature
    (B) extreme weakness
    (C) overactive and hyper-contraction of muscles
    (D) no effect

12. In an action potential, the "threshold" is reached when

    (A) enough voltage-gated channels open to produce a change in membrane polarity
    (B) the synaptic vesicles release neurotransmitters into the synaptic cleft
    (C) the peak depolarization occurs
    (D) sodium ions enter the cell

13. Oligodendricytes exist in the central nervous system, acting analogously to the Schwann cells of the peripheral nervous system. Therefore, their main objective is to

    (A) act as immune cells of the brain
    (B) produce a myelin sheath that insulates the axonal transmission
    (C) direct depolarization of the membrane and perpetuate the nerve impulse
    (D) transmit nerve impulses

14. Testosterone and estradiol are hormones that are derivatives of steroid molecules. Therefore, these hormones reach the interior of many different types of cells easily. Which of the following accurately describes this type of hormone reception?

    (A) Many types of cells have receptors for testosterone and estradiol on their plasma membrane.
    (B) These hormones are lipid-soluble, so they reach the interior of cells by crossing the plasma membrane.
    (C) Transport proteins on the surface of cells shuttle these hormones into the cell.
    (D) Hormones never enter cells, so this question must be inaccurate.

15. Only certain cells in the body are target cells for the hormone vasopressin. Which of the following best explains the specificity of this hormone?

    (A) Target cells are the only cells exposed to vasopressin.
    (B) Target cells are the only cells with receptors for vasopressin.
    (C) Vasopressin enters all cells because it is nonpolar.
    (D) Cells without vasopressin receptors destroy vasopressin before it can have an effect.

REFLECT

Respond to the following questions:

- Which topics from this chapter do you feel you have mastered?

- Which content topics from this chapter do you feel you need to study more before you can answer multiple-choice questions correctly?

- Which content topics from this chapter do you feel you need to study more before you can effectively compose a free response?

- Was there any content that you need to ask your teacher or another person about?

Chapter 12
Behavior and Ecology

In the preceding chapters, we looked at individual organisms and the ways they solve life's many problems: acquiring nutrition, reproducing, and so on. Now let's turn to how organisms deal with their environments. We can divide the discussion of organisms and their environments into two general categories: **behavior** and **ecology**.

BEHAVIOR

Some animals behave in a programmed way to specific stimuli, while others behave according to some type of learning. We humans can do both. For the AP Biology Exam, you'll have to know a little about the different types of behavior.

All behavior means, basically, is how organisms cope with their environments. This chapter looks at the general types of behavior: instinct, imprinting, classical conditioning, and operant conditioning. Let's start with instinct.

Instinct

Instinct is an inborn, unlearned behavior. Sometimes the instinctive behavior is triggered by environmental signals called releasers. The releaser is usually a small part of the environment that is perceived. For example, when a male European robin sees another male robin, the sight of a tuft of red feathers on the male is a releaser that triggers fighting behavior. In fact, because instinct underlies all other behavior, it can be thought of as the circuitry that guides behavior.

For example, hive insects, such as bees and termites, never learn their roles; they are born knowing them. On the basis of this inborn knowledge, or instinct, they carry out all their other behaviors: a worker carries out "worker tasks," a drone "drone tasks," and a queen "queen tasks." Another example of instinct is the "dance" of the honey bee, which is used to communicate the location of food to other members of the beehive.

There are other types of instinct that last for only a part of an animal's life and are gradually replaced by "learned" behavior. For example, human infants are born with an ability to suck from a nipple. If it were not for this instinctual behavior, the infant would starve. Ultimately, however, the infant will move beyond this instinct and learn to feed itself. What exactly, then, is instinct?

For our purposes:

Instinct is the inherited "circuitry" that directs and guides behavior.

A particular type of innate behavior is a **fixed action pattern**. These behaviors are not simple reflexes, and yet they are not conscious decisions. An example is the egg-rolling behavior exhibited by a graylag goose. If the egg is removed from the goose, it will continue to make the same movements. That is, the innate movements are independent of the environment.

Learning

Another form of behavior is **learning**. Learning refers to a change in a behavior brought about by an experience (which is what you're doing this very moment). Animals learn in a number of ways. For the AP Biology Exam, we'll take a look at the key types of learning: imprinting, classical conditioning, operant conditioning, and habituation.

Imprinting

Have you ever seen a group of goslings waddling along after their mother? How is it that they recognize her? Well, the mother arrives and gives out a call that the goslings recognize. The goslings, hearing the call, know that this is their mother and follow her around until they are big enough to head out on their own.

Now imagine the same goslings, newly hatched. If the mother is absent, they will accept the first moving object they see as their mother. This process is known as **imprinting**.

Animals undergo imprinting within a few days after birth in order to recognize members of their own species. While there are different types of imprinting, including parent, sexual, and song imprinting, they all occur during a **critical period**—a window of time when the animal is sensitive to certain aspects of the environment.

Remember:

> Imprinting is a form of learning that occurs during a brief period of time, usually early in an organism's life.

Classical Conditioning

If you have a dog or a cat, you know that every time you hit the electric can opener, your cat or dog comes running. This is a form of **classical conditioning**. To feed your pet, you need to open its can of food. For your pet, the sound of the opener has come to be associated with eating: every time it hears the opener, it thinks that it is about to be fed. We can say that your pet has been "conditioned" to link the buzz of the can opener with its evening meal.

This type of learning is now known as **associative learning**.

Operant Conditioning

Another type of associative learning is **operant conditioning** (or **trial-and-error learning**).

In operant conditioning, an animal learns to perform an act in order to receive a reward.

If the behavior is not reinforced, the conditioned response will be lost. This is called extinction. Here's one thing to remember:

> In operant conditioning, the animal's behavior determines whether it gets the reward or the punishment. Classical conditioning is just the learned association of two things.

Habituation

Habituation is another form of learning. It occurs when an animal learns not to respond to a stimulus. For example, if an animal encounters a stimulus over and over again without any consequences, the response to it will gradually lessen and may altogether disappear.

To recap, there are four basic types of learning:

- Imprinting occurs early in life and helps organisms recognize members of their own species.
- Classical conditioning involves learning through association.
- Operant conditioning occurs when a response is associated with new stimuli (also a form of associative learning).
- Habituation occurs when an animal learns to ignore a stimulus.

Internal Clocks: The Circadian Rhythm

There are other instinctual behaviors that occur in both animals and plants. One such behavior deals with time. Have you ever wondered how roosters always know when to start crowing? The first thought that comes to mind is that they've caught a glimpse of the sun. Yet many crow even before the sun has risen.

Roosters do have internal alarm clocks. Plants have them as well. These internal clocks, or cycles, are known as **circadian rhythms**. Circadian is like the word circle. Hence, the circular nature of it.

Watch out though: seasonal changes, like the loss of leaves by deciduous trees or the hibernation of mammals, are not examples of circadian rhythms. *Circadian* refers only to daily rhythms. Need a mnemonic? Just think how bad your jet lag would be after a trip around the world. In other words: Circling the globe screws up your circadian rhythm.

The Science Behind Your Jet Lag

If you've ever flown overseas, you know all about circadian rhythms. They're the cause of jet lag. Our bodies tell us it's one time, while our watches tell us it's another. The sun may be up, but our body's internal clock is crying, "Sleep!" This sense of time is purely instinctual: you don't need to know how to tell time in order to feel jet lag.

HOW ANIMALS COMMUNICATE

Some animals use signals as a way of communicating with members of their species. These signals, which can be chemical, visual, electrical or tactile, are often used to influence mating and social behavior.

Chemical signals are one of the most common forms of communication among animals. **Pheromones**, for instance, are chemical signals between members of the same species that stimulate olfactory receptors and ultimately affect behavior. For example, when female insects give off their pheromones, they attract males from great distances.

Visual signals also play an important role in the behavior observed among members of a species. For example, fireflies produce pulsed flashes that can be seen by other fireflies far away. The flashes are sexual displays that help male and female fireflies identify and locate each other in the dark.

Other animals use electrical channels to communicate. For example, some fish generate and receive weak electrical fields. Others use tactile signals and have mechanoreceptors in their skin to detect prey. For instance, cave-dwelling fishes use mechanoreceptors in their skin for communicating with other members and detecting prey.

Social Behavior

Many animals are highly social species, and they interact with each other in complex ways. Social behaviors can help members of the species survive and reproduce more successfully. Several behavioral patterns for animal societies are summarized below:

- **Agonistic behavior** is aggressive behavior that occurs as a result of competition for food or other resources. Animals will show aggression toward other members that tend to use the same resources. An example of agonistic behavior is fighting amongst competitors.

- **Dominance hierarchies** (or pecking orders) occur when members in a group have established which members are the most dominant. The more dominant male will often become the leader of the group and will usually have the best pickings of the food and females in the group. Once the dominance hierarchy is established, competition and tension within the group is reduced.

- **Territoriality** is a common behavior when food and nesting sites are in short supply. Usually, the male of the species will establish and defend his territory (called a home range) within a group in order to protect important resources. This behavior is typically found among birds.

- **Altruistic behavior** is defined as unselfish behavior that benefits another organism in the group at the individual's expense because it advances the genes of the group. For example, when ground squirrels give warning calls to alert other squirrels of the presence of a predator, the calling squirrel puts itself at risk of being found by the predator.

Symbiotic Relationships

Many organisms that coexist exhibit some type of **symbiotic relationship.** These include remoras, or "sucker fish," which attach themselves to the backs of sharks, and lichen—the fuzzy, mold-like stuff that grows on rocks. Lichen appears to be one organism, when in fact it is two organisms—a fungus and an alga, or photosynthetic bacterium—living in a complex symbiotic relationship.

Overall, there are three basic types of symbiotic relationships:

- **Mutualism**—in which both organisms win (for example, the lichen components)
- **Commensalism**—in which one organism lives off another with no harm to the host organism (for example, the remora)
- **Parasitism**—in which the organism actually harms its host

There are many special types of relationships between multicellular organisms and unicellular organisms. One important example is bacteria living inside the guts of many mammals. We have a mutualistic relationship. Humans and other mammals provide them a nice home with nutrients, and they help us in many ways by breaking things down and helping us make things. It is also good that we are filled with "good" bacteria because otherwise we might be susceptible to dangerous pathogenic bacteria.

PLANT BEHAVIOR

Plants have also evolved specific ways to respond to their environment. The plant behaviors covered on the AP Biology Exam are photoperiodism and tropisms. Plants flower in response to changes in the amount of daylight and darkness they receive. This is called **photoperiodism**. Although you'd think that plants bloom based on the amount of sunlight they receive, they actually flower according to the amount of uninterrupted darkness.

Tropisms

Plants need light. Notice that all the plants in your house tip toward the windows. This movement of plants toward the light is known as phototropism. As you know, plants generally grow up and down: the branches grow upward, while the roots grow downward into the soil, seeking water. This tendency to grow toward or away from the earth is called gravitropism. All of these tropisms are examples of behavior in plants.

A **tropism** is a turning in response to a stimulus.

There are three basic tropisms in plants. You can remember them because their prefixes represent the stimuli to which plants react:

- **Phototropism** refers to how plants respond to sunlight—for example, bending toward light.

- **Gravitropism** refers to how plants respond to gravity. Stems exhibit negative gravitropism (they grow up, away from the pull of gravity), whereas roots exhibit positive gravitropism (they grow downward into the earth).
- **Thigmotropism** refers to how plants respond to touch. For example, ivy grows around a post or trellis.

ECOLOGY

The study of the interactions between living things and their environments is known as ecology. We've spent most of our time discussing individual organisms. However, in the real world, organisms are in constant interaction with other organisms and the environment. The best way for us to understand the various levels of ecology is to progress from the big picture, the biosphere, down to the smallest ecological unit, the population.

There is a hierarchy within the world of ecology. Each of the following terms represents a different level of ecological interaction:

- **Biosphere**: The entire part of the earth where living things exist. This includes soil, water, light, and air. In comparison to the overall mass of the earth, the biosphere is relatively small. If you think of the earth as a basketball, the biosphere is equivalent to a coat of paint over its surface
- **Ecosystem**: The interaction of living and nonliving things
- **Community**: A group of populations interacting in the same area
- **Population**: A group of individuals that belong to the same species and that are interbreeding

Biosphere

The biosphere can be divided into large regions called **biomes**. Biomes are massive areas that are classified mostly on the basis of their climates and plant life. Because of the different climates and terrains on the earth, the distribution of living organisms varies.

Remember that biomes tend to be arranged along particular latitudes. For instance, if you hiked from Alaska to Kansas, you would pass through the following biomes: tundra, taiga, temperate deciduous forests, and grasslands.

Ecosystem

Ecosystems are self-contained regions that include both living and nonliving factors. For example, a lake, its surrounding forest, the atmosphere above it, and all the organisms that live in or feed off of the lake would be considered an ecosystem. Living factors are called **biotic factors** and nonliving factors are called **abiotic factors.** Abiotic factors include water, humidity, temperature, soil/atmosphere composition, light, and radiation. As you probably know, there is an exchange of materials between the components of an ecosystem.

Take a look at the flow of carbon through a typical ecosystem.

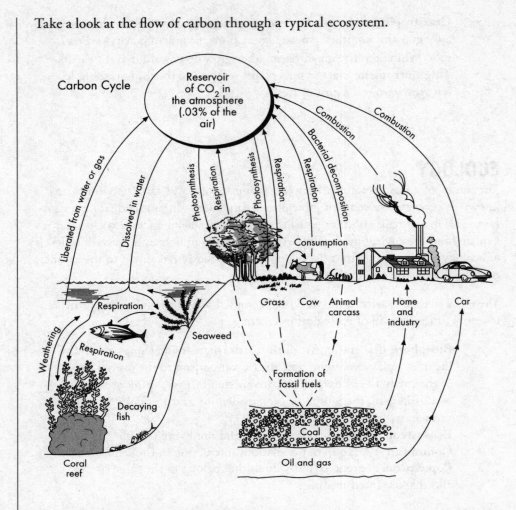

You'll notice how carbon is recycled throughout the ecosystem—this is called the **carbon cycle**. In other words, carbon flows through ecosystems. There are also cycles for nitrogen and water. The interactions between the biotic and the abiotic factors are crucial for the balance of the ecosystem.

Community

The next smaller level is the community. A community refers to a group of inter-acting plants and animals that show some degree of interdependence. For instance, you, your dog, and the fleas on your dog are all members of the same community.

Each organism has its own **niche**—its position or function in a community. When two organisms occupy the same niche, they will compete for the resources within that niche. If a species can occupy an unoccupied niche, it will usually thrive without competition. These connections are shown in the **food chain**. A food chain describes the way different organisms depend on one another for food. There are basically four levels to the food chain: producers, primary consumers, secondary consumers, and tertiary consumers.

Producers (autotrophs)

Producers, or autotrophs, have all of the raw building blocks to make their own food. From water and the gases that abound in the atmosphere and with the aid of the sun's energy, autotrophs convert light energy to chemical energy. They accomplish this through photosynthesis.

Primary productivity is the rate at which autotrophs convert light energy into chemical energy in an ecosystem. There are two types of **primary productivity:**

1. The gross productivity just from photosynthesis cannot be measured because cell respiration is occurring at the same time.

2. Net productivity measures only organic materials that are left over after photosynthetic organisms have taken care of their own cellular energy needs. This is calculated by measuring oxygen production in the light, when both photosynthesis and cell respiration are occurring. This equation is listed on the Biology Equations and Formulas sheet.

Autotrophs produce all of the available food. They make up the first trophic (feeding) level. They possess the highest **biomass** (the total weight of all the organisms in an area) and the greatest numbers. Did you know that plants make up about 99 percent of the earth's total biomass?

Consumers (heterotrophs)

Consumers, or heterotrophs, are forced to find their energy sources in the outside world. Basically, heterotrophs digest the carbohydrates of their prey into carbon, hydrogen, and oxygen and use these molecules to make organic substances.

The bottom line is that heterotrophs, or consumers, get their energy from the things they consume.

Primary consumers are organisms that directly feed on producers. A good example is a cow. These organisms are also known as **herbivores**. They make up the second trophic level.

The next level consists of organisms that feed on primary consumers. They are the **secondary consumers**, and they make up the third trophic level. Above these are **tertiary consumers**.

Decomposers

All organisms at some point must finally yield to decomposers. Decomposers are the organisms that break down organic matter into simple products. Generally, fungi and bacteria are the decomposers. They serve as the "garbage collectors" in our environment.

- Producers make their own food.

- Primary consumers (herbivores) eat producers.

- Secondary consumers (carnivores and omnivores) eat producers and primary consumers.

- Tertiary consumers eat all of the above.

- Decomposers break things down.

Keystone Species

Sometimes one organism is particularly important to an ecosystem. Maybe it is the only producer or maybe it is the only thing that keeps a particularly deadly predator in check. Important species like this are called **keystone species**. If a keystone species is removed from the ecosystem, the whole balance can be undone very quickly.

The 10% Rule

Not For Individuals
This idea about energy does not apply to individuals. It means the level as a whole has less energy if you add up all the members at that level.

In a food chain, only about 10 percent of the energy is transferred from one level to the next—this is called the **10% rule**. The other 90 percent is used for things like respiration, digestion, running away from predators—in other words, it's used to power the organism doing the eating! As a level, the producers have the most energy in an ecosystem; the primary consumers have less energy than producers; the secondary consumers have less energy than the primary consumers; and the tertiary consumers will have the least energy of all.

The energy flow, biomass, and numbers of members within an ecosystem can be represented in an **ecological pyramid**. Organisms that are higher up on the pyramid have less biomass and energy, and they are fewer in number.

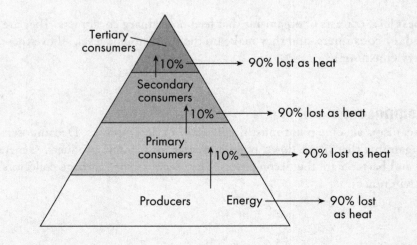

Buildup of Toxins

The downside of food pyramids is that when the consumers at the top eat something beneath them, it is like they are eating that thing AND the thing that thing ate AND and the thing that thing ate, etc. So, if there is a toxin in the environment, the consumers at the top are getting the most of it because it becomes more concentrated at each level.

> Toxins in an ecosystem are more concentrated and thus more dangerous for animals further up the pyramid.

This simply means that if a toxin is introduced into an ecosystem, the animals most likely to be affected are those at the top of the pyramid. This occurs because of the increasing concentration of such toxins. The classic example of this phenomenon is DDT, an insecticide initially used to kill mosquitoes. Although the large-scale spraying of DDT resulted in a decrease in the mosquito population, it also wound up killing off ospreys.

Ospreys are aquatic birds whose diet consists primarily of fish. The fish that ospreys consumed had, in turn, been feeding on contaminated insects (**bioaccumulation**). Because fish eat thousands of insects, and ospreys hundreds of fish, the toxins grew increasingly concentrated (**biomagnification**). Though the insecticide seemed harmless enough, it resulted in the near extinction of certain osprey populations. What no one knew was that in sufficient concentrations, DDT weakened the eggshells of ospreys. Consequently, eggs broke before they could hatch, killing the unborn ospreys.

Such environmental tragedies still occur. All ecosystems, small or great, are intricately woven, and any change in one level invariably results in major changes at all the other levels.

Disrupting the Balance

Ecosystems are never exactly stable, existing in a precarious balance. Many things can disrupt the balance. If the biomass at any level changes, this will affect the surrounding levels and spread throughout the ecosystem. Geological changes, weather changes, new species, diseases, lack of resources, new resources, and many other things can disrupt the balance.

Ecosystem Interconnectedness

Because all levels of an ecosystem are closely connected, a disruption in one level due to environmental changes, for example, affects all other aspects of that ecosystem.

POPULATION ECOLOGY

Population ecology is the study of how populations change. Whether these changes are long-term or short-term, predictable or unpredictable, we're talking about the growth and distribution patterns of a population.

When studying a population, you need to examine four things: the size (the total number of individuals), the density (the number of individuals per area), the distribution patterns (how individuals in a population are spread out), and the age structure.

For example, the graph below shows there is a high death rate among the young of oysters, but those that survive do well. On the other hand, there is a low death rate among the young of humans, but, after age 60, the death rate is high.

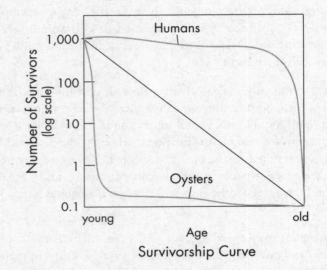

Survivorship Curve

The growth of a population can be represented as the number of births minus the number of deaths divided by the size of the population.

$$r = (births - deaths)/N$$

(r is the reproductive rate, and N is the population size.)

Population growth can also be calculated in the following way:

$$\frac{Change\ in\ Population\ Size}{Change\ in\ Time} = Birth\ Rate - Death\ Rate$$

Each population has a **carrying capacity**—the maximum number of individuals of a species that a habitat can support. Most populations, however, don't reach their carrying capacity because they're exposed to limiting factors.

One important factor is **population density**. The factors that limit a population are either density-independent or density-dependent. **Density-independent factors** affect the population, regardless of the density of the population. Some examples are severe storms and extreme climates. On the other hand, **density-dependent**

factors are those with effects that depend on population density. Resource depletion, competition, and predation are all examples of density-dependent factors. In fact, these effects become even more intense as the population density increases.

Exponential Growth

The growth rates of populations also vary greatly. There are two types of growth: exponential growth and logistic growth. Equations for both are included on the AP Biology Equations and Formulas sheet. **Exponential growth** occurs when a population is in an ideal environment. Growth is unrestricted because there are lots of resources, space, and no disease or predation. Here's an example of exponential growth. Notice that the curve arches sharply upward—the exponential increase.

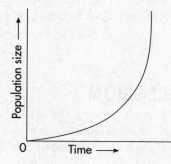

Exponential growth occurs very quickly, resulting in a J-shaped curve. A good example of exponential growth is the initial growth of bacteria in a culture. There's plenty of room and food, so they multiply rapidly.

However, as the bacterial population increases, the individual bacteria begin to compete with each other for resources. When the resources become limited, the population reaches its carrying capacity, and the curve levels off.

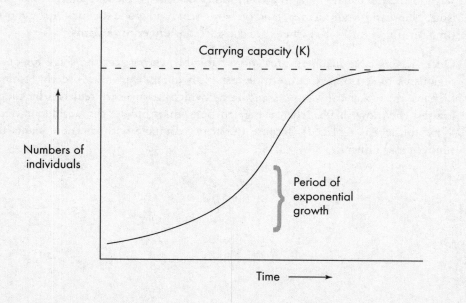

The population becomes restricted in size because of limited resources. This is referred to as **logistic growth**. Notice that the growth forms an S-shaped curve.

These growth patterns are associated with two kinds of life-history strategies: r-selected species and k-selected species.

R-strategists tend to thrive in areas that are barren or uninhabited. Once they colonize an area, they reproduce as quickly as possible. Why? They know they've got to multiply before competitors arrive on the scene! The best way to ensure their survival is to produce lots of offspring. Typical examples are common weeds, dandelions, and bacteria.

At the other end of the spectrum are the **k-strategists**. These organisms are best suited for survival in stable environments. They tend to be large animals, such as elephants, with long lifespans. Unlike r-strategists, they produce few offspring. Given their size, k-strategists usually don't have to contend with competition from other organisms.

ECOLOGICAL SUCCESSION

Communities of organisms don't just spring up on their own; they develop gradually over time. **Ecological succession** refers to the predictable procession of plant communities over a relatively short period of time (decades or centuries).

Centuries may not seem like a short time to us, but if you consider the enormous stretches of time over which evolution occurs, hundreds of thousands or even millions of years, you'll see that it is pretty short.

The process of ecological succession in which no previous organisms have existed is called **primary succession**.

How does a new habitat full of bare rocks eventually turn into a forest? The first stage of the job usually falls to a community of lichens. Lichens are hardy organisms. They can invade an area, land on bare rocks, and erode the rock surface, over time turning it into soil. Lichens are considered **pioneer organisms**.

Once lichens have made an area more habitable, they've set the stage for other organisms to settle in. Communities establish themselves in an orderly fashion. Lichens are replaced by mosses and ferns, which in turn are replaced by tough grasses, then low shrubs, then evergreen trees, and, finally, the deciduous trees. Why are lichens replaced? Because they can't compete with the new plants for sunlight and minerals.

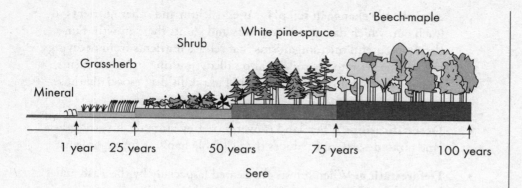

The entire sequence is called a **sere**. The final community is called the **climax community**. The climax community is the most stable. In our example, the deciduous trees are part of the climax community.

Now what happens when a forest is devastated by fire? The same principles apply, but the events occur much more rapidly. The only exception is that the first invaders are usually not lichens, but grasses, shrubs, saplings, and weeds. When a new community develops where another community has been destroyed or disrupted, this event is called **secondary succession**.

HUMAN IMPACT ON THE ENVIRONMENT

Unfortunately, humans have disturbed the existing ecological balance, and the results are far-reaching. Soils have been eroded and various forms of pollution have increased. The potential consequences on the environment are summarized below.

- **Greenhouse effect:** The atmospheric concentrations of carbon dioxide increased by the burning of fossil fuels and forests have contributed to the warming of the earth. Higher temperatures may cause the polar ice caps to melt and flooding to occur. Other potential effects of global warming include changes in precipitation patterns, changes in plant and animal populations, and detrimental changes in agriculture.

- **Ozone depletion:** Pollution has also led to the depletion of the atmospheric ozone layer by such chemicals as chlorofluorocarbons (CFCs), which are used in aerosol cans. Ozone (O_3) forms when UV radiation reacts with O_2. Ozone protects the earth's surface from excessive ultraviolet radiation. Its loss could have major genetic effects and could increase the incidence of cancer.

- **Acid rain:** Burning fossil fuels produces pollutants such as sulfur dioxide and nitrogen dioxide. When these compounds react with droplets of atmospheric water in clouds, they form sulfuric and nitric acids, respectively. The rain that falls from these clouds is weakly acidic and is called acid rain. Acid rain lowers the pH of aquatic ecosystems and soil, which damages water systems, plants, and soil. For example, the change in soil pH causes calcium and other nutrients to leach out, which damages plant roots and stunts their growth. Furthermore, useful microorganisms that release nutrients from decaying organic matter into the soil are also killed, resulting in less nutrients being available for the plants. Low pH also kills fish, especially those that have just hatched.

- **Desertification:** When land is overgrazed by animals, it turns grasslands into deserts and reduces the available habitats for organisms.

- **Deforestation:** When forests are cleared (especially by the slash and burn method), erosion, floods, and changes in weather patterns can occur.

- **Pollution:** Another environmental concern is the toxic chemicals in our environment. One example is DDT, a pesticide used to control insects. DDT was overused at one time and later found to damage plants and animals worldwide. DDT is particularly harmful because it resists chemical breakdown, and it can still be found in the tissues of nearly every living organism. The danger with toxins such as DDT is that as each trophic level consumes DDT, the substance becomes more concentrated by a process called **biomagnification**.

- **Reduction in biodiversity:** As different habitats have been destroyed, many plants and animals have become extinct. Some of these plants could have provided us with medicines and products that may have been beneficial.

Acid rain, deforestation, pollution, soil erosion, and depletion of the ozone layer are just a few of the problems plaguing the environment today.

KEY TERMS

behavior
ecology
instinct
fixed action pattern
learning
imprinting
critical period
classical conditioning
associative learning
operant conditioning
   (or trial-and-error learning)
habituation
circadian rhythms
pheromones
agonistic behavior
dominance hierarchy
territoriality
altruistic behavior
symbiotic relationship
mutualism
commensalism
parasitism
photoperiodism
tropism
phototropism
gravitropism
thigmotropism
biosphere
ecosystem
community
population
biomes
biotic factors
abiotic factors
carbon cycle
niche
food chain

producers (autotrophs)
primary productivity
biomass
consumers (heterotrophs)
primary consumers
herbivores
secondary consumers
tertiary consumers
decomposer
keystone species
10% rule
ecological pyramid
bioaccumulation
biomagnification
population growth
carrying capacity
population density
density-independent factors
density-dependent factors
exponential growth
logistic growth
r-strategists
k-strategists
ecological succession
primary succession
pioneer organisms
sere
climax community
secondary succession
greenhouse effect
ozone depletion
acid rain
desertification
deforestation
pollution
biomagnification
reduction in biodiversity

Summary

Behavior

o   Behavior is an organism's response to the environment. Behavior can be instinctual (inborn), learned, or social:
  - Types of learned behaviors include imprinting, classical conditioning, and operant conditioning.
  - Types of social behaviors include agonistic behavior, dominance hierarchies, territoriality, and altruistic behavior.

o   There are also plant-specific behaviors known as tropisms. The three basic tropisms are phototropism, gravitropism, and thigmotropism.

Ecology

o   There are several major biomes (tundra, taiga, temperate deciduous forest, grasslands, deserts, tropical rainforests) that make up the biosphere. Each biosphere contains ecosystems.

o   Within an ecosystem are communities, which consist of organisms fulfilling one of three main roles:
  - Producers, or autotrophs, convert light energy to chemical energy via photosynthesis.
  - Consumers, or heterotrophs, acquire energy from the things they consume. Their digestion of carbohydrates produces carbon, hydrogen, and oxygen, which are then used to make organic substances.
  - Decomposers form fossil fuels from the detritus of other organisms in the ecosystem.
  - The 10% rule says that only 10% of the energy consumed from one level will be retained by the higher level that consumed it. The other energy will be spent to perform normal daily activities.

o   The smallest unit of ecology is the population. The growth of a population can be found with the equation $(r) = (births - deaths) / N$.

o   The carrying capacity is the maximum number of individuals that can be supported by a habitat. Most populations do not reach carrying capacity due to factors such as population density (density-independent factors and density-dependent factors).

o   Exponential growth (J-shaped curve) occurs when a population is an ideal environment. Logistic growth (S-shaped curve) of a population occurs when there are limited resources in an environment.

o   Organisms are generally either r-strategists or k-strategists. R-strategists ensure their survival by

producing lots of offspring. K-strategists, which are usually large animals, produce few offspring but have longer life spans and less competition from other organisms for resources.

○ Succession describes how ecosystems recover after a disturbance in terms of pioneers, sere, climax community, and secondary succession.

○ Human impact on the planet includes the following issues:
  • greenhouse effect
  • ozone depletion
  • acid rain
  • desertification
  • deforestation
  • pollution
  • reduction in biodiversity

Chapter 12 Drill

Answers and explanations can be found in Chapter 16.

1. A chimpanzee stacks a series of boxes on top of one another to reach a bunch of bananas suspended from the ceiling. This is an example of which of the following behaviors?

   (A) Operant learning
   (B) Imprinting
   (C) Instinct
   (D) Insight

2. Viruses are obligate intracellular parasites, requiring their host cells for replication. Consequently, viruses generally attempt to reproduce as efficiently and quickly as possible in a host. Below is a graph depicting the initial growth pattern of a bacteriophage within a population of *E. coli*. This reproductive strategy is most similar to which of the following?

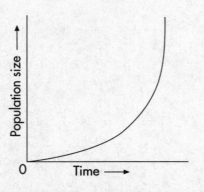

   (A) An r-strategist, because it aims to produce a large abundance of offspring to ensure survival
   (B) A k-strategist, because it aims to produce a large abundance of offspring to ensure survival
   (C) An r-strategist, because it is best suited to thrive in stable environments and over a long life-span
   (D) A k-strategist, because it is best suited to thrive in stable environments and over a long life-span

3. In a pond ecosystem, spring rains trigger an expansion of species at levels of the food chain. Runoff from nearby hills brings nutrients which, when combined with warming temperatures, trigger an algae bloom. The populations of small protozoans such as plankton expand by ingesting the algae. These plankton, subsequently, are consumed by small crustaceans such as crayfish, which ultimately become prey for fish such as catfish or carp. In this ecosystem, which of the following accurately describes the crayfish?

   (A) They are producers.
   (B) They are primary consumers.
   (C) They are secondary consumers.
   (D) They are tertiary consumers.

Questions 4-6 refer to the following table, chart, and paragraph.

Table 1: Minimum pH Tolerance of Common Aquatic Organisms

Animal	pH Minimum
Brook Trout	4.9
American Bullfrog	3.8
Yellow Perch	4.6
Tiger Salamander	4.9
Crayfish	5.4
Snails	6.1
Clams	6.0

Chart 1: Change in Populations of Aquatic Organisms in Richard Creek (1980–2000)

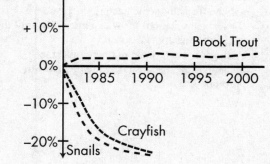

Over the period of 1980 to 2000, the average pH in Richard Creek changed drastically. An ecological survey was performed to evaluate the effect of detectable decreases in pH on the aquatic life of the creek. Four times a year, ecological surveys were performed to identify the number of snails (a primary consumer), crayfish (a secondary consumer), and brook trout (a tertiary consumer) present at five different locations. The percent change relative to 1980 is shown in **Chart 1** above. Many aquatic organisms cannot live in low pH conditions. The minimum pH necessary for common aquatic organisms to sustain life is shown in **Table 1**.

4. According to the data in Chart 1 and Table 1, the average pH in the creek is most nearly which of the following?

(A) 5.9
(B) 5.5
(C) 5.1
(D) 4.7

5. Which of the following is the most likely explanation for the abrupt decrease in pH in Richard Creek?

(A) Greenhouse effect
(B) Deforestation
(C) Acid rain
(D) Pollution

6. In order for snails to return to Richard Creek, the pH of the creek must exceed

(A) 5.4
(B) 5.7
(C) 6.0
(D) 6.1

7. *Mimosa pudica* is a plant often called the "sensitive plant" because when you touch the leaves, they immediately close up. One theory about the purpose of this type of movement is that herbivores avoid the plant due to this movement. The movement of *Mimosa pudica* is an example of

(A) phototropism
(B) gravitropism
(C) aquatropism
(D) thigmotropism

Question 8 refers to the following graph.

Foodprints by Diet Type: t CO_2^e/person

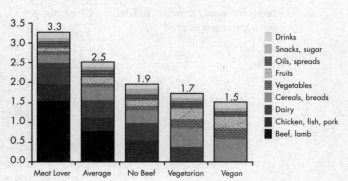

8. A proponent of the vegan diet writes a blog post that claims that the vegan diet is the only way to "reduce your carbon footprint, with respect to food, because it cuts your carbon footprint in half" and shows the graph above. Which of the following directly supports this claim?

(A) According to the 10% rule, only 10% of energy is passed from producers to consumers, so when humans consume animals, the energy harvested is greatly reduced.
(B) Livestock produce methane emissions and consume large amounts of plants every year.
(C) Deforestation for grazing livestock contributes to the loss of plants for carbon sequestration.
(D) Beef and lamb livestock—not chicken, fish, or pork—has the most dramatic impact on carbon footprint.

9. Costa Rica is plentiful with many beautiful tropical rainforests. For a long time, cattle farmers cleared much of the forests for their livestock. However, in recent decades, the government has placed more emphasis on reforestation. Several fields are now being reclaimed by the rainforest. One such field has many long grasses, a few tall palms, and some dense shrubbery along the periphery of the field. This field can be said to be

(A) invaded by pioneer species
(B) a product of secondary succession
(C) a climax community
(D) filled with invasive species

10. Probiotics are beneficial bacteria or yeast applied to maintain intestinal health in humans or livestock. In theory, these organisms inhabit the intestines and aid in digestion of food. Also, these organisms have bioprotective effects. How can addition of bacteria to the intestine protect the host from disease?

(A) Probiotic bacteria crowd the intestine, preventing pathogenic bacteria from growing due to density-dependent limitation on their population.
(B) Probiotic bacteria crowd the intestine, preparing a niche for pathogenic bacteria to grow.
(C) Bacteria populations grow exponentially until the carrying capacity is reached.
(D) Pathogenic bacteria do not recognize the intestinal cells because they are coated with probiotic bacteria.

REFLECT

Respond to the following questions:

- Which topics from this chapter do you feel you have mastered?

- Which content topics from this chapter do you feel you need to study more before you can answer multiple-choice questions correctly?

- Which content topics from this chapter do you feel you need to study more before you can effectively compose a free response?

- Was there any content that you need to ask your teacher or another person about?

Chapter 13
Quantitative Skills and Biostatistics

Don't forget to check out
the formula sheet!

There will be six designated quantitative questions on the exam, but this does not mean that the multiple-choice and the free-response questions will not be quantitative. These types of questions pop up all over the place. Even if you think you are a whiz, check out the formula pages at the back of this book. Expect at least a couple questions using formulas from these sheets.

It is often easy to see patterns in data, but it is not always easy to determine if a pattern is valid or significant. **Quantitative data** analysis is the first step in figuring this out. In this chapter, we will review how data can be summarized, presented, and tested for validity. This depends on what kind of question was being asked at the beginning of the experiment. When you're reading this chapter, pay attention to which techniques are used when. On the AP Biology Exam, you should be able to justify which techniques are best.

This list is a good reference for the types of quantitative questions you should be able to tackle. We will not walk though all of these in detail, but you will find some examples in the end-of-chapter drill and in the practice tests. Use the formula sheets to familiarize yourself with these concepts.

Must-Know Types of Quantitative Problems

- ☐ Hardy-Weinberg Equilibrium
- ☐ Water Potential/Osmosis
- ☐ Energy Pyramids/Biomass
- ☐ Chi-squared Analysis
- ☐ Gene Linkage Analysis
- ☐ Inheritance I Probability
- ☐ Reading from Graphs
- ☐ Predicting from Graphs
- ☐ Gibbs Free Energy
- ☐ Dilutions of Solutions
- ☐ Population Growth
- ☐ Primary Productivity
- ☐ Q_{10}

SUMMARIZING AND PRESENTING DATA

Instead of presenting raw data, scientists use **descriptive statistics** and graphs to summarize large datasets, present patterns in data, and communicate results. Descriptive statistics summarize and show variation in the data. There are many types of descriptive statistics, including measures of central tendency (such as mean, median, and mode) and measures of variability (such as standard deviation, standard error, and range). The AP Biology Equations and Formulas sheet contains definitions for mean, median, mode, and range. It also includes equations for mean, standard deviation, and standard error. That said, you will not need to calculate standard deviation or standard error. Instead, focus on understanding how these values are used.

Descriptive statistics are used to summarize the data collected from samples, but may also describe the entire population you're trying to study. Experiments are designed to include a **sample**, which is a subset of the **population** being studied. The best experiments use random sampling, which makes sure there is no bias when picking which individuals from the population will be included in the sample. It is important that the sample is big enough that the data from the experiment are a good representation of what would happen if the whole population were measured.

Graphs are visual representations of data and are often used to reveal trends that might not be obvious by looking at a table of numbers. There are six types of graphs you should be familiar with:

1. Bar graph
2. Pie graph
3. Histogram
4. Line graph
5. Box-and-whisker plot
6. Scatterplot

Each is described below. Pay attention to which graph is used when. On the AP Biology Exam, you may need to decide which graph is best and then create it. You may also have to analyze graphs given to you on the exam.

A good graph must include the following things:

*   a good title that describes what the experiment was and what was measured
*   axes labeled with numbers, labels and units, and index marks
*   a frame or perimeter
*   data points that are clearly marked (e.g., easily identifiable lines or bars)

TYPES OF DATA

Data can be **quantitative** (based on numbers or amounts that can be measured or counted) or **qualitative** (data that is descriptive, subjective, or difficult to measure). For example, behavioral observations are qualitative. Most data on the AP Biology Exam will be quantitative, with either counts or measurements.

Count Data

Count data are generated by counting how many of an item fit into a category. For example, you could count the number of organisms with a particular phenotype or the number of animals in one habitat versus another. Count data also include data that is collected as percentages or the results of a genetic cross. This type of data is usually summarized using a **bar graph** or a **pie graph**. For example, suppose a mixed population of *E. coli* were plated on growth plates. Some cells contained a β-galactosidase marker and grew in blue colonies, while others did not contain the marker and grew in white colonies. The colonies were counted after 24 hours of growth.

E.coli Colony Growth on Plates after 24 hours

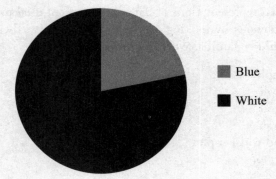

In another example, suppose fly populations were monitored in northern Maine to determine if population size varies over the warm months of the year.

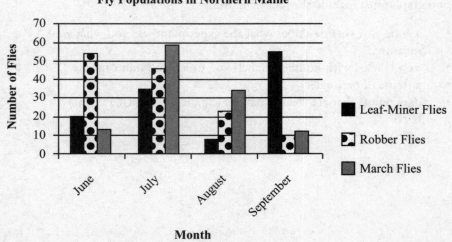

Count data can be analyzed using hypothesis testing, which will be described later in this chapter.

Measurements

Measurements are continuous, meaning there is an infinite number of potential measurements over a given range. Size, height, temperature, weight, and response rate are all measurements. There are two types of measurement data: parametric and nonparametric.

Normal or Parametric Data

Normal or parametric data are measurement data that fit a **normal curve or distribution**, usually when a large sample size is used. For example, if you took a large sample of 17-year-olds in America and graphed the frequency of heights, the results will be normally distributed.

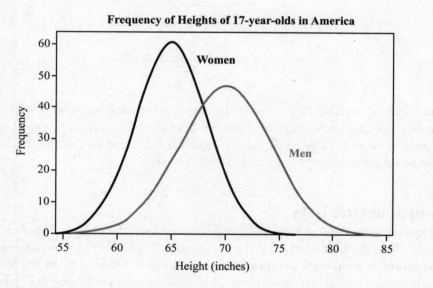

Frequency of Heights of 17-year-olds in America

Several descriptive statistics can be used to summarize normal data.

The **sample size** (*n*) refers to the number of members of the population that are included in the study. Sample size is an important consideration when you're trying to determine how well the data in the study represents a population. Large sample sizes are always better, but for technical reasons, experiments can't be infinitely large.

The **mean** ($\overline{x}$) is the average of the sample, calculated by adding all of the individual values and dividing by the number of values you have. The mean is not necessarily a number provided in the sample. You should be able to recognize what the mean of a given dataset is and be able to calculate the mean using the equation on the AP Biology Equations and Formulas sheet. One limitation of mean is that it is influenced by **outliers**, or numerical observations that are far removed from the rest of the observations.

The **standard deviation** (*s*) can determine if numbers are packed together or dispersed because it is a measure of how much each individual number differs from the mean. A low standard deviation means the data points are all similar and close to the mean, while a high standard deviation means the data are more spread out. Here is the relationship between a normal distribution and standard deviation (or SD).

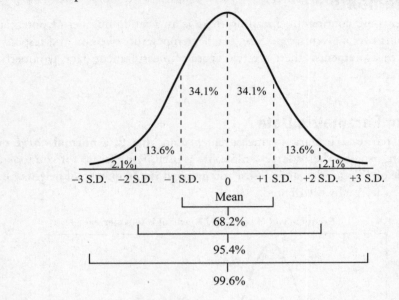

This means that about 70 percent of the data is within one standard deviation of the mean in a normally distributed dataset. **Standard error** (*SE*) can also be used to report how much a given dataset varies and is calculated by dividing the standard deviation by the square root of the sample size.

Nonparametric Data

Nonparametric data often include large outliers and do not fit a normal distribution. In order to determine if data is parametric or not, you can construct a **histogram** or **frequency diagram**. These graphs give information on the spread of the data and the central tendencies. Making a histogram is like setting up bins, or intervals with the same range, that cover the entire dataset (the *x*-axis). The measurements in each bin are graphed on the *y*-axis. For example, suppose you counted the number of pine needles on each branch of a *Pinus strobus* tree. A histogram of this data may look like this.

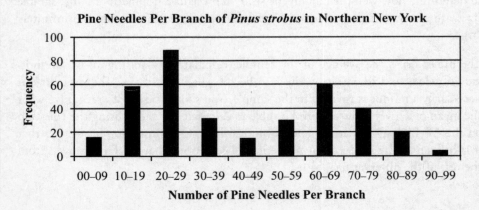

Pine Needles Per Branch of *Pinus strobus* in Northern New York

This dataset does not match a normal distribution, meaning it is a nonparametric dataset. Keep in mind that even if a certain dataset is not normal, the population itself could be. There could have been sampling bias or errors in the data collection, which can lead to sample data that doesn't match population data. The best way to fix this is usually to increase the sample size. Histograms differ from bar graphs in that they show ranges instead of just categories.

Nonparametric data requires a different set of descriptive statistics than parametric data.

The **median** is the middle number in a dataset and is determined by putting the numbers in consecutive order and finding the middle number. If there is an odd number of numbers, there will be a single number that is the median. If there is an even number of numbers, the median is determined by averaging the two middle numbers. Therefore, the median is not necessarily one of the numbers in the dataset. The median is useful in gauging the midpoint of the data, but will not necessarily tell you much about outliers.

The **mode** is the most frequently recurring number in the dataset. If there are no numbers that occur more than once, there is no mode. If there are multiple numbers that occur most frequently, each of those numbers is a mode. The mode must be one of the numbers in the sample, and modes are never averaged.

The **range** is the difference between the smallest and largest number in a sample. This value is less useful than standard deviation, because it does not give any information on individual values or the majority of values.

Example 1: Eleven male laboratory mice were weighed at 32 weeks of age. The data collected was: 32 g, 28 g, 29 g, 34 g, 30 g, 28 g, 32 g, 31 g, 30 g, 32 g, 33 g. What is the mean, median and mode of this dataset?

Solution: Let's start by putting the 11 numbers in the dataset in order, from smallest value to largest value: 28 g, 28 g, 29 g, 30 g, 30 g, 31 g, 32 g, 32 g, 32 g, 33 g, 34 g. The mean is

$$Mean = \frac{28 + 28 + 29 + 30 + 30 + 31 + 32 + 32 + 32 + 33 + 34}{11}$$

$$= \frac{339}{11}$$

$$= 30.8\,g$$

The median is the middle value, or 31 g. The mode is the most frequent number, or 32 g.

Example 2: In the following set of values, what is the range?

Values: −5, 8, 11, −1, 0, 4, 14

Solution: The smallest value in the set above is −5, and the largest is 14. The difference between these two is the range, which is 19.

TYPES OF EXPERIMENTS OR QUESTIONS

You will be tested on scientific experiments and the scientific reasoning behind different aspects of the experiments a LOT. You should be clear with the steps of the scientific method and know how to properly design an experiment and recognize a poorly designed experiment. The following list should remind you:

- **Hypothesis:** A well-thought-out prediction of what the outcome of the experiment will be.
- **Independent Variable:** The factor that you, as the experimenter, will change between the different groups in the experiment. If one plant is grown in pH 5 and the other in pH 7, then pH is the independent variable.
- **Dependent Variable:** The data. The thing that you measure during the experiment. The height of the plant you are growing might be the dependent variable.
- **Constants (Controlled Variables):** The things that are the same between all your groups. Hint: Everything should be constant except the independent variable.
- **Control Groups:** Any group that is needed simply so you can compare your interesting experimental groups to it. These are often the "no treatment" group.
- **Statistical Significance:** This is how trustworthy the results are and how sure you can be in your conclusions about them. This can always be increased by including more individuals in the groups or including more trials.

Most biological experiments do one of three things:

1. look at how something changes over time
2. compare groups of some sort
3. test for an association

How you present and summarize data depends on what kind of experiment you perform.

Time-Course Experiments

Time-course experiments look at how something changes over time. A **line graph** is usually used to present this type of data. Several lines can be plotted on the same graph, but these must be clearly labeled. Also, each dot on a line graph could represent one data point or a mean of values. If mean values are plotted, standard deviation or standard error can also be shown. More generally, line graphs can also be used to compare how a dependent variable (*y*-axis) changes with an independent variable (*x*-axis).

When making line graphs, the intervals on the axes must be consistent. Also make sure it is clear whether the data start at the origin (0, 0) or not. The same guidelines apply to scatterplots, which will be discussed below.

Data points should be connected with a solid line. You can extrapolate (or continue) the line past the data points, but then must use a broken line. Finally, the slope on a line graph tells you the rate of change.

Suppose we want to determine how stress hormone levels change after the end of an important exam. If we measure cortisol and epinephrine in the serum of several people every hour for five hours after the end of an exam, we would have several values for each time point. We could plot the mean of each measurement and also show data variance using standard error bars. In this example, the distance above the data point is the standard error, and the distance below the data point is the standard error. We call this "±SE."

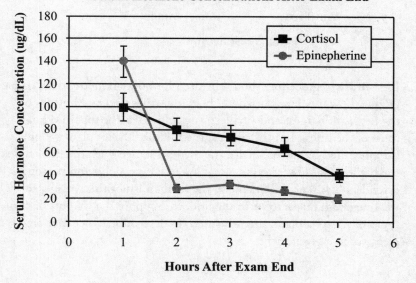

Comparative Experiments

These experiments compare populations, groups, or events. There are several options for graphing data from these experiments:

- **Bar graphs** are helpful to compare categories of data if the data is parametric. It is also a good idea to show variance of the data by including the standard error. This gives information on how different the two means are from each other. Suppose you compare two strains of fission yeast before and after treatment with a drug that inhibits cell wall synthesis. As a measure of viability, you measure how much Myc mRNA was made in each strain with and without treatment. In this case, each bar is a mean of values and the error bars are ±SE, as described above.

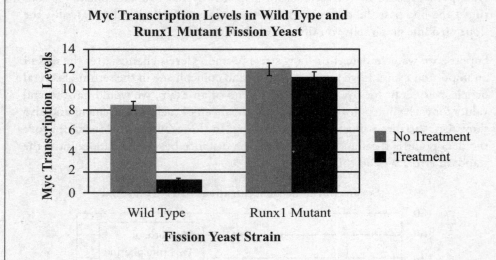

Myc Transcription Levels in Wild Type and Runx1 Mutant Fission Yeast

- **Box-and-whisker plots** should be used for nonparametric data. These graphs show median (horizontal line), quartiles (top and bottom of the box), and largest and smallest values (ticks at the tops and bottoms of the lines). In other words, the middle 50% of plant weights is represented by the box, with the median shown with the line, and the highest and lowest values are marked with the top and bottom extenders (whiskers). Remember, the median is not the average, so the line doesn't have to be in the middle. Suppose the weights of two types of Arizona plants are measured, to determine which is more plentiful in a given niche:

Grass and Cactus Weights in the Arizona Desert

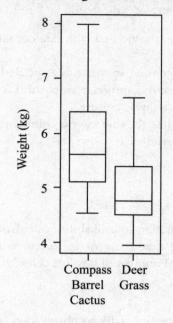

In order to conclude that the samples or groups being compared are different or not, hypothesis testing should be performed. This will be discussed later in this chapter.

Association Experiments

These experiments look for associations between variables. They attempt to determine if two variables are correlated, and additional tests can demonstrate causation. **Scatterplots** are used to present data from association experiments. Each data point is plotted as a dot. Suppose different substrate concentrations are used in an enzymatic reaction (using the enzyme Kozmase III), and the enzyme efficiency is measured (in percent of maximum). The data could be presented like this.

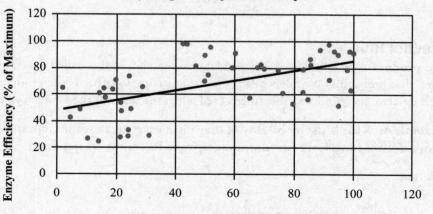

Enzyme Efficiency and Substrate Concentration for Kozmase III at Physiological Temperature and pH

If the relationship looks linear, a linear regression line can be added. There are a few common shapes of scatterplots you should be familiar with:

- A bell-shaped curve is associated with random samples and normal distributions.
- An upward curve (concave in shape) is associated with exponentially increasing functions. A common example of this is the lag phase of bacterial growth in a liquid culture.
- A curve that looks like the sine wave is common when biological rhythms are being studied.

PROBABILITY

You already saw an introduction to probability calculations in Chapter 8 of this book. The **probability** (P) that an event will occur is the number of favorable cases (a) divided by the total number of possible cases (n).

$$P = a / n$$

This can be determined experimentally by observation (such as when population data is being collected) or by the nature of the event. For example, the probability of getting a two when rolling a die is 1/6, since there are six sides on a die.

Example 3: What is the probability of drawing the nine of hearts from a deck of cards?

Solution: Since there are 52 cards in a deck and this question is asking about one particular card, $P = 1/52$.

Combining Probabilities

Many genetics problems involve several probabilities, all being considered together to answer a larger biological question. There are three rules that are often used.

Product Rule

The **product rule** is used for independent events and is also called the "AND rule." The probability of independent events occurring together is the product of their separate probabilities. This rule is used when the order of the events matters.

Example 4: What is the probability of drawing a heart and a spade consecutively from a deck of cards if the first draw was replaced before the second draw?

Solution:

$$P(\text{heart}) = \frac{13}{52} = \frac{1}{4}$$

$$P(\text{spade}) = \frac{13}{52} = \frac{1}{4}$$

$$P(\text{heart and spade}) = P(\text{heart}) \times P(\text{spade})$$

$$P(\text{heart and spade}) = \frac{1}{4} \times \frac{1}{4} = \frac{1}{16}$$

Sum Rule

The **sum rule** is used when studying two mutually exclusive events, and can be thought of as the "EITHER/OR" rule." The probability of either of two events occurring is the sum of their individual probabilities. This rule is used when the order of the events matters.

Example 5: What is the probability of drawing a diamond or a heart from a deck of cards?

Solution:

$$P(\text{diamond}) = \frac{13}{52} = \frac{1}{4}$$

$$P(\text{heart}) = \frac{13}{52} = \frac{1}{4}$$

$$P(\text{diamond or heart}) = P(\text{diamond}) + P(\text{heart})$$

$$P(\text{diamond or heart}) = \frac{1}{4} + \frac{1}{4} = \frac{2}{4} = \frac{1}{2}$$

HYPOTHESIS TESTING

t-test and *p*-values

Many experiments involve comparing two datasets or two groups, and a **t-test** can be used to calculate whether the means of two groups are significantly different from each other. This test is most often applied to datasets that are normally distributed. A **p-value** equal to or below 0.05 is considered significant in most biology-related fields. *T*-test calculations involve comparing each data point to the group's mean and also take standard deviation and sample size into consideration.

Understanding mathematically how *p*-values are calculated and *t*-tests are done is not important here. Instead, it's important that you understand how these values are used to interpret data.

Let's work through an example to demonstrate how common statistics are used in biological labs. A researcher has sections of two different types of skin cancers from human patients. She stains them for the protein CD31, which is a marker of endothelial cells. She then takes digital pictures of the immunofluorescent sections and counts the number of blood vessels present in each picture. She takes five pictures of each slide.

Slide A	Slide B
10	3
8	2
7	4
8	4
11	4

(a) What is the mean, median, and mode for each dataset?

(b) The standard error of group A is 0.735, and the standard error for group B is 0.400. What does this tell you about the data?

(c) What assumptions are made about the data in performing these tests?

(d) What could the researcher do to increase her confidence in the results?

Solutions:

(a) The means are

$$Mean_{Group\ A} = \frac{10 + 8 + 7 + 8 + 11}{5} = 8.8$$

$$Mean_{Group\ B} = \frac{3 + 2 + 4 + 4 + 4}{5} = 3.4$$

Remember, the median is the middle number, so it's best to put the data in order first.

Slide A	Slide B
7	2
8	3
8	4
10	4
11	4

The median of group A is 8, and the median of group B is 4.

The mode is the most frequent value. For group A, this is 8. For group B, the mode is 4.

(b) Standard error is the standard deviation divided by the square root of the sample size. The sample size is five for both group A and group B because five pictures were taken for each slide. Because the standard error of group A is larger than that of group B and because the two groups have the same sample size, the standard deviation of group A must also be larger than that of group B. Note that this may or may not be the case if the sample sizes were different. A larger standard deviation means the data is more variable, so you can conclude that the data from group A has a larger spread than the data from group B.

(c) As with most datasets, the researcher is assuming the data fits a normal distribution and that her sample size is large enough to be meaningful.

(d) More data allows for more confident conclusions. The researcher could therefore take more pictures from each slide to increase the sample size.

Chi-Square (x^2) Tests

A **chi square test** is a statistical tool used to measure the difference between observed and expected data. For example, suppose you roll a die 120 times. You would expect a 1:1:1:1:1:1 ratio between all the numbers that come up. In other words, you would expect to see each side 20 times. However, because this is an experiment and the sample size is only 120, you will probably see some variation in the data. If you rolled 24 sixes for example, the disagreement between observed (24) and expected (20) is small, and could have occurred by chance. If you rolled 40 sixes however, there is a large disagreement from the expected. Maybe it is a weighted die. The region in between these two values is trickier. Where is the line drawn between expected/normal variations and results that are so different that they must tell a different story altogether?

Chi-square tests are one way to make this decision. They start with a **null hypothesis** (H_o), which represents a set numerical outcome that you want to compare your data with. The chi-square test will determine if your actual data is close to the null hypothesis outcome and could have occurred due to a fluke (i.e., rolling a six 24 times) or if it is so different that it is probably not a fluke (i.e., 40 rolled sixes) and the null hypothesis outcome must not correctly describe what is occurring. In the dice case, rolling a six 40 times would cause the null hypothesis of expecting a 1:1:1:1:1:1 to be rejected and the die can be considered to be weighted. In the calculations, expected results if the null hypothesis were true are compared to the actual values, and an x^2 value is calculated. This value is compared to a **critical value**, and a decision is made to reject or accept the null hypothesis.

For example, suppose data was collected from 100 families with three children each. Fourteen families had three female children, 36 families had two females and a male, 30 families had one female and two males, and 20 families had three male children. We would expect offspring sex to segregate independently. In other words, the sex of a given child shouldn't be affected by the sex of previous children. Let's test whether this is true for this dataset.

Since we expect sex to segregate independently, this is the null hypothesis (H_o). Each family has three children, so there are four possible outcomes for each family:

1. three daughters
2. two daughters and a son
3. one daughter and two sons
4. three sons

Using some advanced math (that you don't need to worry about), the probabilities for each of these is:

1. P(three females) = 12.5%
2. P(two females, one male) = 37.5%
3. P(one female, two males) = 37.5%
4. P(three males) = $(1)(0.50)^3$ = 12.5%

Because there are 100 families, we expect 12.5 to have three females, 37.5 to have two females and a male, 37.5 to have one female and two males, and 12.5 to have three males. This can be calculated by multiplying the probability of each event by the total (100 in this case).

Next, we compare this expected data (E) to what actually happened (the observed data, O). Note that the total numbers for observed and expected columns must be the same. To calculate an x^2 value, the formula $(O - E)^2/E$ is calculated for each row of the table, and summed.

Family Type	Observed (O)	Expected (E)	$(O - E)^2/E$
3 females	14	12.5	0.18
2 females 1 male	36	37.5	0.06
1 female 2 males	30	37.5	1.5
3 males	20	12.5	4.5
Total	**100**	**100**	**6.24**

The calculated x^2 value is 6.24. Next, this value needs to be compared to a critical value (CV). This is obtained from a table like this one.

p-value	Chi-Square Table Degrees of Freedom							
	1	2	3	4	5	6	7	8
0.05	3.84	5.99	7.82	9.49	11.07	12.59	14.07	15.51
0.01	6.64	9.21	11.34	13.28	15.09	16.81	18.48	20.09

This chart is included on the AP Biology Equations and Formulas sheet. In order to use this table, you must know the **degrees of freedom** (DF), which represent the number of independent variables in the data. In most cases, this is the number of possibilities being compared minus one. In this example, DF = 4 − 1, because we are comparing four different family types.

You also need a reference p-value. This is the probability of observing a deviation from the expected results due to chance. For most biological tests, a p-value of 0.05 is used.

Since we are using DF = 3 and p = 0.05, the critical value is 7.82 (using the chart above). Next, the critical value is compared to the x^2 value; if x^2 < CV, you accept H_o. If x^2 > CV, you reject H_o. We calculated a x^2 value of 6.24, which is smaller than 7.82. Since x^2 < CV, we accept H_o and conclude that offspring sex could, indeed, be segregating independently in this dataset.

KEY TERMS

descriptive statistics

sample

population

quantitative data

qualitative data

count data

bar graph

pie graph

normal or parametric data

normal curve or distribution

sample size

mean

outliers

standard deviation

standard error

nonparametric data

histogram or frequency diagram

median

mode

range

hypothesis

independent variable

dependent variable

contants (contolled variables)

controls

statistical significance

time-course experiment

line graph

bar graph

box-and-whisker plot

scatterplot

probability

product rule

sum rule

t-test

p-value

chi-square test

null hypothesis

critical value

degrees of freedom

Summary

- Descriptive statistics (such as mean, standard deviation, standard error, median, mode, range) and graphs are used to summarize data and show patterns and conclusions:
  - Normal (parametric) measurement data are summarized using mean and standard deviation or standard error.
  - Parametric data are summarized using median, mode, and range.
  - Histograms can be used to see if a dataset has a standard deviation or not.

- Count data are summarized using a bar graph or a pie graph.

- Time course experiments look at how something changes over time; summarize these using a line graph.

- Experiments that compare groups are summarized using a bar graph (if the dataset has a standard deviation) or a box-and-whisker plot (if it does not).

- Scatterplots summarize association experiments, and regression lines can be used to determine if the relationship is linear.

- Probability can be calculated using the sum rule or the product rule.

- Hypothesis testing is used to determine if two groups are significantly different from each other. They start with a null hypothesis, which is rejected or accepted, depending on how a calculated p-value or chi-square value compares to a standard value.

Chapter 13 Drill

Answers and explanations can be found in Chapter 16.

1. Five subjects were weighed before and after an 8-week exercise program. What is the average amount of weight lost in pounds for all five subjects, rounded to the nearest pound?

Subject	Starting Weight (pounds)	Final Weight (pounds)
1	184	176
2	200	190
3	221	225
4	235	208
5	244	225

(A) 12 pounds
(B) 13 pounds
(C) 14 pounds
(D) 15 pounds

2. The height of six trees is measured. Is plant 6 taller than the median for all six trees?

Plant	Height (inches)
1	67
2	61
3	72
4	71
5	66
6	68

(A) Yes, the median is 67.3.
(B) No, the median is 67.3.
(C) Yes, the median is 67.5.
(D) No, the median is 67.5.

3. In the following set of test scores, what is the mode and what is the range?

Test Scores: 71, 67, 75, 65, 66, 32, 69, 70, 72, 82, 73, 68, 75, 68, 75, 78

(A) Mode: 68; Range: 75
(B) Mode: 69; Range: 50
(C) Mode: 75; Range: 70.5
(D) Mode: 75; Range: 50

4. Given the cross $AaBbCc \times AaBbCc$, what is the probability of having an $AABbCC$ offspring?

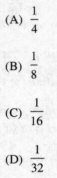

(A) $\dfrac{1}{4}$

(B) $\dfrac{1}{8}$

(C) $\dfrac{1}{16}$

(D) $\dfrac{1}{32}$

5. Given the cross $AaBb \times aabb$, what is the probability of having an $Aabb$ or $aaBb$ offspring?

(A) $\dfrac{1}{2}$

(B) $\dfrac{1}{4}$

(C) $\dfrac{1}{16}$

(D) 0

6. Two pea plants are crossed, and a ratio of 3 yellow plants to 1 green plant is expected in the offspring. It is found that out of 100 plants phenotyped, 84 are yellow and 16 are green. Do the experimental data match the expected data?

(A) Yes, the x^2 value is greater than 3.84.
(B) Yes, the x^2 value is smaller than 3.84.
(C) No, the x^2 value is greater than 3.84.
(D) No, the x^2 value is smaller than 3.84.

7. A mating is set up between two pure breeding strains of plants. One parent has long leaves and long shoots. The other parent has short leaves and stubby shoots. F_1 plants are collected, and all have long leaves and long shoots. F_1 plants are self-crossed, and 1,000 F_2 plants are phenotyped. The data is as follows:

Phenotype	# of F_2
Long leaves, long shoots	382
Long leaves, stubby shoots	109
Short leaves, long shoots	112
Short leaves, stubby shoots	397
Total	1,000

Are the genes for leaf and shoot length segregating independently?

(A) Yes; the degrees of freedom are 3, and the calculated χ^2 value is small.

(B) No; the degrees of freedom are 3, and the calculated χ^2 value is large.

(C) Yes; the degree of freedom is 1, and the calculated χ^2 value is small.

(D) No; the degree of freedom is 1, and the calculated χ^2 value is large.

REFLECT

Respond to the following questions:

- Which topics from this chapter do you feel you have mastered?

- Which content topics from this chapter do you feel you need to study more before you can answer multiple-choice questions correctly?

- Which content topics from this chapter do you feel you need to study more before you can effectively compose a free response?

- Was there any content that you need to ask your teacher or another person about?

Chapter 14
Sample
Free-Response
Questions

SECTION II: FREE RESPONSE

Section II of the AP Biology Exam contains eight questions, consisting of two long free-response questions and six short free-response questions. You will have a total of 90 minutes—including 10 minutes for reading—to complete all questions. The directions will look something like this (read carefully!).

> **Directions:** Questions 1 and 2 are long free-response questions that should require about 22 minutes each to answer and are worth 10 points each. Questions 3 through 8 are short free-response questions that should require about 6 minutes each to answer. Questions 3 through 5 are worth 4 points each, and questions 6 through 8 are worth 3 points each.
>
> Read each question carefully and completely. You are advised to spend the 10-minute reading period planning your answers. You may begin writing your responses before the reading period is over. Write your response in the space provided following each question. Only material written in the space provided will be scored. Answers must be written out in paragraph form. Outlines, bulleted lists, or diagrams alone are not acceptable.

The following sections discuss scoring guidelines in more detail and provide sample questions. But first take some time to review some key tips for writing your free responses.

Free-Response Tips and Strategies

AP Biology Readers (the people who grade your free-response essays) expect students to interpret data and apply knowledge to new situations in the free-response questions. Some important things to keep in mind when writing your answers to the free-response questions are provided in the following list:

- **Answer the question.** Do not become distracted or waste time writing about tangential topics that you know more about than what is being asked. Focus your response.

- **Think quantitatively.** Use data whenever possible to explain your answer. This is especially necessary when the question features a graph, table, or other type of diagram (which many free-response questions now do).

- **Be specific.** Offer specific detailed examples where appropriate.

- **Use vocabulary correctly and often.** Hone your understanding of the words provided in the Key Terms lists in this book and the various biological ideas and situations to which they can be applied. This will help you use the correct terminology when writing your responses.

- **Remember the Big Ideas.** Know the four big ideas covered in the AP Biology course (page 49), as they are reflected in many free-response questions. Understanding these ideas and being able to articulate them will help you craft your responses.

Okay, now it's time for some practice. Take about 22 minutes and try to write a long free-response to the following question. Keep these tips in mind as you write.

The regulation of the expression of gene products is crucial to the development of an embryo and the maintenance of homeostasis. The table below illustrates the mRNA levels, total protein levels, and total enzyme activity levels for three different enzymes. These enzymes are encoded by genes A, B, and C after being expressed in both the absence and presence of "Protein X" or "Protein Y."

	Untreated			+ Protein X			+ Protein Y		
	Gene A	Gene B	Gene C	Gene A	Gene B	Gene C	Gene A	Gene B	Gene C
mRNA Levels	1,200	2,000	1,400	2,100	250	1,400	1,200	2,000	1,400
Protein Levels	400	1,100	300	750	20	300	400	1,100	300
Total Enzyme Activity Levels	20,000	32,000	18,000	35,000	5	18,000	20,000	50,000	200

Describe the likely role of Protein X and/or Protein Y in the expression of each of these genes. **Include** when during gene expression the regulation occurs and **explain** a possible mechanism including a potential binding partner for Protein X and/or Protein Y.

In a follow-up experiment, Protein X and Protein Y are added together. On the axes below, **create** a graph (or several) predicting the outcome on the mRNA, protein, and enzyme activity of Genes A, B, and C. Be sure to show the untreated levels also.

SCORING GUIDELINES

To help you grade this sample free response, we've put together a checklist that you can use to calculate the number of points that should be assigned to each part of this question. We'll first explain the important points on the checklist and then give you sample student responses to show you how test reviewers would evaluate them.

Free-Response Checklist

1. Describe the roles/point of regulation/possible mechanism and binding partner (**2 points for each correct combination for 6 points total**)

    a. gene A regulation (transcription enhanced by X)

    b. gene B regulation (transcription inhibited by X and enzyme activity enhanced by Y)

    c. gene C regulation (enzyme activity inhibited by Y)

2. Create graph(s)

    a. The graph(s) should show the correct trend for each gene in comparison to untreated. (**1 point each for 3 points total**)

        i. gene A (mRNA, protein, and enzyme activity increased)

        ii. gene B (mRNA, protein, and enzyme activity reduced, though maybe not as much of a reduction in enzyme as was seen with protein X alone)

        iii. gene C (mRNA and protein unchanged, but enzyme activity reduced)

    b. The graph should have properly labelled axes (most likely with the levels of mRNA, protein, and enzyme activity on the Y-axis and the different genes on the X-axis). (**1 point**)

SAMPLE ESSAYS

The following response would earn full points.

Gene A expression has no change with the addition of Protein Y, but it has increased transcription, protein, and enzyme activity with the addition of Protein X. It is likely that Protein X is an enhancer of transcription of gene A. Perhaps it binds directly to the promoter region of gene A and assists RNA polymerase to bind. It might also remove a repressor of transcription or work together with other transcription factors. The increase in mRNA would lead to increased protein and increased enzyme activity down the road.

Gene B has a decrease in mRNA, protein, and enzyme activity with Protein X. It has increased total activity with Protein Y. It is likely that Protein X is a repressor of transcription for gene B. It might bind to the promoter of gene

B or to another transcription factor to prevent RNA polymerase binding or prevent transcription from occurring in another way. Since the enzyme activity increases with Protein Y, but the mRNA and the protein levels did not change, this means that the enzyme activity was somehow enhanced on the same amount of enzyme that was present without Protein Y. Protein Y might be an enzyme activator of Enzyme B. It might bind to Enzyme B directly or maybe the protein phosphorylates it or cleaves it to make the enzyme active.

Gene C is unaffected by Protein X, but Protein Y decreases the enzyme activity without affecting the mRNA or protein levels. This means it is acting upon the enzyme and turning it off. Protein Y might be an allosteric or a competitive inhibitor of Enzyme C. If it is an allosteric inhibitor, then it binds to a site other than the active site. If it is a competitive inhibitor, then it binds to the active site on Enzyme C.

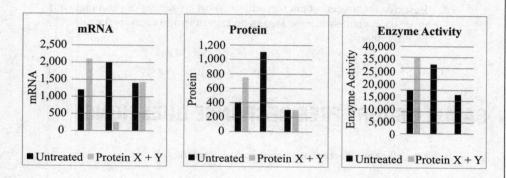

HOW TO USE THE SAMPLE FREE-RESPONSE QUESTIONS

The most important advice for this section of the test is *practice, practice, practice!* No matter how well you think you know the material, it's important to practice formulating your thoughts on paper. Before you take our practice tests in the back of the book, make sure you review our sample free-response questions. Then use our checklist—which is put together just like the testing board's—to assign points.

To give you as much practice as possible, we've compiled a few sample long and short free-response questions. Most of these questions refer to themes in biology that crop up time and time again on the exam, so chances are you'll see something similar on your test. You can also access free-response questions from recent exams on the College Board website at apstudent.collegeboard.org/apcourse/ap-biology/exam-practice.

SAMPLE LONG FREE-RESPONSE QUESTIONS

1. **Describe** how the modern theory of evolution is supported by evidence from the following areas:

   (a) Comparative anatomy
   (b) Biogeography
   (c) Embryology

2. **Describe** the chemical nature of genes. **Discuss** the replicative process of DNA in eukaryotic organisms. What are the various types of gene mutations that can occur during replication?

3. **Describe** the process of photosynthesis. What are the major plant pigments involved in photosynthesis? **Design** an experiment to measure the rate of photosynthesis.

SAMPLE SHORT FREE-RESPONSE QUESTIONS

1. **Describe** the structure of a generalized eukaryotic cell.

2. **Discuss** the following responses in plants and give one example of each.

   (a) Photoperiodism
   (b) Phototropism

3. **Select** one of the following pairs of hormones and **discuss** the concept of negative feedback.

   (a) Thyroid-stimulating hormone (TSH) and thyroxine
   (b) Parathyroid hormone and calcitonin
   (c) Cortisol and ACTH

The biology knowledge you need to answer all of these free-response questions is in this book. After you have completed your essays, go back to the chapters that cover these topics and check to see if you have presented all the relevant points. Use our Free-Response Checklist as a guideline to give yourself (or have someone give you) an approximate score. Good luck!

Summary

- The free-response section features eight questions, two of which require long responses. You will have 90 minutes to complete the entire section.

- Each question is worth a certain number of points. Generally, the long-response questions are worth 10 points, and the short-response questions are worth either 3 or 4 points each.

- Your responses must be written in paragraph form; outlines are not acceptable. Diagrams can be used to supplement your writing but will not receive credit on their own (unless otherwise specified).

- Remember to keep your response focused and use specific data and examples (where applicable) to justify your answers.

- Familiarize yourself with the key terms in this book as well as the Big Ideas for the AP Biology course. Understanding these will help you write clear, focused, and accurate responses.

REFLECT

Respond to the following questions:

- Which topics from this chapter do you feel you have mastered?

- Which content topics from this chapter do you feel you need to study more before you can answer multiple-choice questions correctly?

- Which content topics from this chapter do you feel you need to study more before you can effectively compose a free response?

- Was there any content that you need to ask your teacher or another person about?

Chapter 15
Laboratory

All AP Biology courses have a laboratory component that gives students hands-on experience regarding some of the biology topics covered in class. Through these laboratory exercises, you can learn the scientific method, lab techniques, and problem-solving skills. There is not an official set of labs that must be completed, and AP teachers get to choose the ones for their classroom. This means that you will not be tested on memorizing specific protocols or data from the labs. However, all labs teach the same basic skills of critical thinking, developing hypotheses, judging results, making conclusions, adapting experiments, and understanding variables. So, those are the skills that you are expected to fully understand. The exam will likely present you with example experiments and you will need to make decisions and answer questions about them. There are 13 labs that are very popular to do in AP Biology classes, and these are the same labs that often come up on the AP test. We have given you short summaries of those 13 labs. This way, if you encounter a question about one of these labs, you will be one step ahead of students that have never even heard of it before.

LAB 1: ARTIFICIAL SELECTION

This lab explores artificial selection—the process by which humans decide traits to enhance or diminish in other species by crossing individuals with the desired phenotype. You need to understand the basic principles of natural selection and how natural selection drives evolution, listed below:

- Variation is present in any population.

- Natural selection, or differential reproduction in a population, is a major mechanism in evolution.

- Natural selection acts on phenotypic variations in populations. Some organisms will have traits that are more favorable to the environment and will survive to reproduce more than other individuals, causing the genetic makeup of the population to change over time.

This lab specifically deals with Wisconsin Fast Plants. In order to artificially select for certain traits, plants with the desired traits can be crossed, changing the genetic makeup of the population. For example, if you wanted to select for height, you could cross only the tallest plants with one another. The new population will have a higher mean height than the previous generation.

LAB 2: MATHEMATICAL MODELING: HARDY-WEINBERG

In order to understand everything you need for the AP Biology Exam, you need to be able to use the Hardy-Weinberg principles and equations to determine allele frequencies in a population. One way to study evolution is to study how the frequencies of alleles change from generation to generation:

- Know how to calculate the allele and genotype frequencies using the two Hardy-Weinberg equations: $p + q = 1$ and $p^2 + 2pq + q^2 = 1$. Don't forget: if the population obeys Hardy-Weinberg's rules, these frequencies remain constant over time.

- Review the discussion of the Hardy-Weinberg principle in this book. Know the five conditions of the Hardy-Weinberg equilibrium: (1) large population, (2) no mutations, (3) no immigration or emigration, (4) random mating, and (5) no natural selection.

- Review natural selection and how it can lead to changes in the genetic makeup of a population.

LAB 3: COMPARING DNA SEQUENCES TO UNDERSTAND EVOLUTIONARY RELATIONSHIPS WITH BLAST

This laboratory uses BLAST (Basic Local Alignment Search Tool), a database that allows you to input a DNA sequence for a gene to look for similar or identical sequences present in other species. In order to understand everything you need for the AP Exam you need to be able to do the following:

- Look at a phylogenic tree, which is a way to visually represent evolutionary relatedness. Endpoints of each branch correspond to a specific species, and each junction on the tree represents a common ancestor. Species that are closer on a phylogenic tree are more closely related.

- Understand BLAST scoring. Most phylogenic trees are constructed by examining nucleotide sequences; the more identical two species sequences are for a specific gene, the more closely related they are. BLAST is able to analyze different sequences to tell you how similar they are to one another based on a score. The higher the score, the closer the two sequences align.

LAB 4: DIFFUSION AND OSMOSIS

This lab investigates the process of diffusion and osmosis in a semipermeable membrane as well as the effects of solute concentration and water potential on these processes.

What are the general concepts you really need to know?

Fortunately, this lab covers the same concepts about diffusion and osmosis that are discussed in this book. Just remember that osmosis is the movement of water across a semipermeable membrane, from a region of high water concentration to one of low water concentration, or from a hypotonic region (low solute concentration) to a hypertonic region (high solute concentration).

- Be familiar with the concept of water potential, which is simply the free energy of water. It is a measure of the tendency of water to diffuse across a membrane. Water moves across a selectively permeable membrane from an area of higher water potential to an area of lower water potential.

- Be familiar with the effects of water gain in animal and plant cells. In animals, the direction of osmosis depends on the concentration of solutes both inside and outside of the cell. In plants, osmosis is also influenced by turgor pressure—the pressure that develops as water presses against a cell wall. If a plant cell loses water, the cell will shrink away from the cell wall and plasmolyze.

Another important concept to understand is the importance of surface area and volume in cells. There are several questions about each of these topics, but the necessary formulas are listed on the AP Biology Equations and Formulas sheet. Cells maintain homeostasis by regulating the movement of solutes across the cell membrane. Small cells have a large surface area-to-volume ratio; however, as cells become larger, this ratio becomes smaller, giving the cell relatively less surface area to exchange solutes. A cell is limited in size by the surface area-to-volume ratio. There are many organisms that have evolved strategies for increasing surface area, like root hairs on plants and villi in the small intestines of animals.

LAB 5: PHOTOSYNTHESIS

The chemical equation for photosynthesis is

$$6CO_2 + 6H_2O \rightarrow C_6H_{12}O_6 + 6O_2$$

Because plants consume some of this energy during photosynthesis, measuring the oxygen produced by a plant can tell us about the net photosynthesis that is occurring. In this laboratory, photosynthesis rates are measured using leaf discs that begin to float as photosynthesis is carried out, allowing you to see that photosynthesis is occurring.

There are several properties that affect the rates of photosynthesis, including:

- light intensity, color, and direction
- temperature
- leaf color, size, and type

Be able to hypothesize about the effects of these variables: for example, as light intensity increases, so does the rate of photosynthesis. Remember, both plants and animals contain mitochondria and carry out cell respiration!

LAB 6: CELLULAR RESPIRATION

In this lab the respiratory rate of germinating and nongerminating seeds and small insects is investigated. The equation for cellular respiration is

$$C_6H_{12}O_6 + O_2 \rightarrow 6CO_2 + 6H_2O$$

Germinating seeds respire and need to consume oxygen in order to continue to grow. Non-germinating seeds do not respire actively. In this lab, the amount of oxygen consumed by these types of seeds is measured with a respirometer. The experiment is also conducted at two temperatures, 25°C and 10°C, because seeds consume more oxygen at higher temperatures. You should know the following:

- Oxygen is consumed in cellular respiration.

- Germinating seeds have a higher respiratory rate than non-germinating seeds, which have a very low respiratory rate.

- Know how to design a study to determine the effect of temperature on cell respiration.

- Know the significance of a control using glass beads. A control is a condition held constant during an experiment. In this case, glass beads are used as a control because they will not consume any oxygen.

LAB 7: MITOSIS AND MEIOSIS

This lab highlights the differences between mitosis and meiosis. In this lab, slides of onion root tips are prepared to study plant mitosis. The important information and skills to review in this lab include the following:

- Mitosis produces two genetically identical cells, while meiosis produces haploid gametes.

- Cell division is highly regulated by checkpoints that depend, in part, on complexes of proteins called cyclins with other proteins called cyclin-dependent kinases. One such example is mitosis promoting factor (or MPF) that has its highest concentration during cell division and is thought to usher a cell into mitosis.

- Nondisjunction, or the failure of chromosomes to separate correctly, can lead to an incorrect number of chromosomes (too many or too few) in daughter cells.

- Know what each phase of the cell cycle looks like under a microscope. Chapter 8 contains diagrams of each phase.

In one section of this lab, the sexual life cycle of the fungus Sordaria fimicola is examined. Sexual reproduction in this fungus involves the fusion of two nuclei: a (+) strain and a (−) strain—to form a diploid zygote. This zygote immediately undergoes meiosis to produce asci, which contain eight haploid spores each.

- During meiosis, crossing-over can occur to increase genetic variation. If crossing-over, or recombination, has occurred, different genetic combinations will be observed in the offspring when compared with the parent strain.

- These offspring with new genetic combinations are called recombinants. By examining the numbers of recombinants with the total number of offspring, an estimate of the linkage map distance between two genes can be calculated using the following equation.

Map distance (in map units) = [(# recombinants) / (# total offspring)] × 100.

LAB 8: BIOTECHNOLOGY: BACTERIAL TRANSFORMATION

In this lab, the principles of genetic engineering are studied. Biotechnologists are able to insert genes into an organism's DNA in order to introduce new traits or phenotypes, like inserting genes into a corn genome that help the crops ward off pests. This process is very complicated in higher plants and animals, but relatively simple in bacteria. You are responsible for knowing the ways in which bacteria can accept fragments of foreign DNA:

- Conjugation: the transfer of genetic material between bacteria via a pilus, a bridge between the two cells

- Transformation: when bacteria take up foreign genetic material from the environment

- Transduction: when a bacteriophage (a virus that infects bacteria) transfers genetic material from one bacteria to another

In addition, DNA can also be inserted into bacteria using plasmids, which are small, circular DNA fragments that can serve as a vector to incorporate genes into the host's chromosome. Plasmids are key elements in genetic engineering. The concepts you need to know about plasmids include the following:

- One way to incorporate specific genes into a plasmid is to use restriction enzymes, which cut foreign DNA at specific sites, producing DNA fragments. A specific fragment can be mixed together with a plasmid, and this recombinant plasmid can then be taken up by *E. coli*.

- Plasmids can give a transformed cell selective advantage. For example, if a plasmid carries genes that confer resistance to an antibiotic like ampicillin, it can transfer these genes to the bacteria. These bacteria are then said to be transformed. That means if ampicillin is in the culture, only transformed cells will grow. This is a clever way scientists can find out which bacteria have taken up a plasmid.

- In order to make a bacteria cell take up a plasmid, you must (1) add $CaCl_2$, (2) heat shock the cells, and (3) incubate them in order to allow the plasmid to cross the plasma membrane.

LAB 9: BIOTECHNOLOGY: RESTRICTION ENZYME ANALYSIS OF DNA

This laboratory introduces you to the technique of electrophoresis. This technique is used in genetic engineering to separate and identify DNA fragments. You need to know the steps of this lab technique for the AP exam:

1. DNA is cut with various restriction enzymes.

2. The DNA fragments are loaded into wells on an agarose gel.

3. As electricity runs through the gel, the fragments move according to their molecular weights. DNA is a negatively charged molecule; therefore, it will migrate toward the positive electrode. The longer the DNA fragment, the slower it moves through the gel.

4. The distance that each fragment has traveled is recorded.

Restriction mapping allows scientists to distinguish between the DNA of different individuals. Since a restriction enzyme will cut only a specific DNA sequence, it will cause a unique set of fragments for each individual called restriction fragments length polymorphisms, or RFLPs for short. This technology is used at crime scenes to help match DNA samples to suspects.

LAB 10: ENERGY DYNAMICS

This lab examines energy storage and transfer in ecosystems. Almost all organisms receive energy from the sun either directly or indirectly:

- Producers, or autotrophs, are organisms that can make their own food using energy captured from the sun. They convert this into chemical energy that is stored in high-energy molecules like glucose.

- Consumers, or heterotrophs, must obtain their energy from organic molecules in their environment. They can then take this energy to make the organic molecules they need to survive.

- Biomass is a measure of the mass of living matter in an environment. It can be used to estimate the energy present in an environment.

LAB 11: TRANSPIRATION

This lab investigates the mechanisms of transpiration, the movement of water from a plant to the atmosphere through evaporation. What do you need to take away from this lab?

- The special properties of water that allow it to move through a plant from the roots to the leaves include polarity, hydrogen bonding, cohesion, and adhesion.

- Know the vascular tissues that are involved in transport in plants. Xylem transports water from roots to leaves, while phloem transports sugars made by photosynthesis in the leaves down to the stem and roots.

- Stomata are small pores present in leaves that allow CO_2 to enter for photosynthesis and are also a major place where water can exit a plant during transpiration.

LAB 12: FRUIT FLY BEHAVIOR

In this lab, fruit flies are given the choice between two environments using a choice chamber, which allows fruit flies to move freely between the two environments. Typically, fruit flies prefer an environment that provides either food or a place to reproduce. They also respond to light and gravity:

- Taxis is the innate movement of an organism based on some sort of stimulus. Movement toward a stimulus is called positive taxis, while movement away from a stimulus is negative taxis. In this lab, fruit flies exhibited a negative gravitaxis (or a movement opposite to the force of gravity) and positive phototaxis (a movement toward a light source).

LAB 13: ENZYME ACTIVITY

This lab demonstrates how an enzyme catalyzes a reaction and what can influence rates of catalysis. In this lab the enzyme peroxidase is used to catalyze the conversion of hydrogen peroxide to water and oxygen.

$$H_2O_2 + \text{peroxidase} \rightarrow 2H_2O + O_2 + \text{peroxidase}$$

The following are the major concepts you need to understand for the AP exam:

- Enzymes are proteins that increase the rate of biological reactions. They accomplish this by lowering the activation energy of the reaction.

- Enzymes have active sites, which are pockets that the substrates (reactants) can enter that are specific to one substrate or set of substrates.

- Enzymes have optimal temperature and pH ranges at which they catalyze reactions. Enzyme concentration and substrate concentration can also influence the rates of reaction.

- If you're asked to design an experiment to measure the effect of these four variables on enzyme activity, keep all the conditions constant except for the variable of interest. For example, to measure the effects of pH in an experiment, maintain the temperature, enzyme concentration, and substrate concentration constant as you change pH.

REFLECT

Respond to the following questions:

- Which topics from this chapter do you feel you have mastered?

- Which content topics from this chapter do you feel you need to study more before you can answer multiple-choice questions correctly?

- Which content topics from this chapter do you feel you need to study more before you can effectively compose a free response?

- Was there any content that you need to ask your teacher or another person about?

Chapter 16
Chapter Drill
Answers and
Explanations

CHAPTER 4 DRILL: CHEMISTRY OF LIFE

1. **C** Water has many unique properties that favor life, including (A), a high specific heat, (B), high surface tension and cohesive properties, and (D), high intermolecular forces due to hydrogen bonding. However, water is a very polar molecule and is an excellent solvent, making (C) inaccurate and the correct answer.

2. **B** The pH scale is logarithmically based, meaning that each difference of 1 on the log scale is indicative of a ten-fold difference in the hydrogen (H^+) ion concentration. According to the question, the pH values of the cytoplasm of cells and gastric juices are approximately 7.4 and 2.5, respectively. Therefore, the pH values vary by 5, and the hydrogen (H^+) ion concentrations much vary by 5 orders of ten (10^5), or 100,000-fold. Since lower pH values have more hydrogen ions, they are also more acidic.

3. **C** All amino acids share a carboxylic acid group, COOH, labeled (B), and an amino-group, NH_2, labeled (D). They also share a hydrogen atom bound to the central carbon (A). Differences in amino acids are defined by variations in the fourth position called the R group, labeled (C).

4. **B** The hypothesis that life may have arisen from formation of complex molecules from the primordial "soup" of Earth is not supported by the absence of nucleic acids. All life is DNA-based, yet no nucleic acid molecules were detected. The presence of carbon molecules, amino acids, and sugars, which are common compounds and compose life, supports the hypothesis, so eliminate (A), (C), and (D). Choice (B) is correct.

5. **C** The amino acids cysteine and methionine contain the element sulfur (as indicated by the *S* in the amino acid structure shown). However, no sulfur-based compounds were included in the Miller-Urey experiment, so it was impossible to form these two amino acids under the conditions of the experiment.

6. **B** Silica is a mineral form of glass, is not a common component of life-forms, and is largely chemically inert. Since oxygen is already present in several compounds included in the experiment, the addition of this compound does not provide any additional elements or chemical substrates, which would permit generation of additional amino acids or synthesis of nucleic acids. The addition of sulfur compounds and phosphorus are necessary to generate some amino acids and all nucleic acids, which eliminates (A), (C), and (D).

7. **D** Hydrolysis adds water to a polymer to break the linkages between monomers. Therefore, (A), free water, will not result because water is being broken down in the reaction. Free water would result from the opposite reaction, for condensation reactions. Choice (B), adenine, results from the hydrolysis of DNA, as it is a monomer of DNA. Choice (C), cholesterol, is a steroid and therefore does not undergo hydrolysis. Dipeptides could result from the hydrolysis of proteins, which are composed of polypeptides.

8. **B** Phospholipids have a long fatty acid tail and a polar head group. The fatty acid tails associate in the membrane, and the phospholipid head group associates with water at the boundary of the membrane. Nucleotides, (A), are mostly polar and found in DNA and RNA. Water, (C), is polar and not found in plasma membranes. Amino acids, (D), are zwitterionic, that is, they have both a positive and a negative charge. Free amino acids are not found in the membrane.

9. **C** $(CH_2O)_3 = C_{18}H_{36}O_{18}$, as in (A), but the dehydration reactions to produce two glycosidic bonds between the three monosaccharides would remove two H_2O molecules, or four hydrogens and two oxygens. Therefore, the correct answer is (C).

10. **A** All isotopes of hydrogen will contain one proton, by definition. All atoms that have only one proton are identified as hydrogen. Tritium is an isotope that has one proton and two neutrons, for a total of three particles in its nucleus. Choice (A) is correct. Choice (B) is not correct because two protons would mean that the element is helium, not hydrogen. Choice (C) is not correct because the atomic number, one, will never change without changing the identity of the atom. Finally, (D) is not correct because radioactive atoms do not give off electrons.

CHAPTER 5 DRILL: CELLS

1. **C** Active transport is the movement of substances across a membrane against their concentration gradient through the use of energy (ATP). The sodium-potassium pump is a critical structure which uses ATP hydrolysis to move sodium and potassium ions against their respective concentration gradients. Diffusion of oxygen is an example of simple diffusion and can occur without the need of channels or pores. Water uses aquaporins to travel across the membrane, down its concentration gradient; therefore, it also does not require energy. Movement of sodium ions by a voltage-gated ion channel, (B), is an example of facilitated diffusion. Because the sodium ions are still undergoing diffusion from high to low concentrations without the need of ATP, this does not represent a form of active transport.

2. **C** Bacterial cells can be visualized using light microscopy. In fact, back in the 17th century, some of the earliest studies using primitive microscopes recorded the shape and organization of bacteria. Choices (A), (B), and (D) are incorrect because virus and cell organelle structures are too small to be observed using light microscopy and require electron microscopy.

3. **D** The cell wall is a structure that is present in bacteria but absent in animal cells. Consequently, this structure is targeted by several leading classes of antibiotics and would be an effective target of therapeutics against *V. cholerae*. Cytoplasm, plasma membrane, and ribosomes—(A), (B), and (C), respectively—are all structures that are present in animal cells.

4. **C** Based on the microscopy data, the organism has a cell wall and lacks mitochondria. The presence of a cell wall suggests that the organism is not a protozoan, so eliminate (B). The absence of mitochondria is indicative of prokaryotic structure because they lack organelles, eliminating (A) and (D). The most likely conclusion is that this organism is a bacterial species.

5. **700** The graphed line of chlorophyll *a* has its highest peak at 700 nm.

6. **B** Although many of the mentioned organelles work closely together, the only choice that can be correct is (B). Ribosomes translate and manufacture proteins, often on the rough endoplasmic reticulum. Those proteins are then packaged in the Golgi bodies into vesicles, which are then fused with the plasma membrane. Choice (A) is incorrect because the nuclear envelope and nucleolus do not interact with vacuoles; they interact with the centrioles only during mitosis. Choice (C) is incorrect because ribosomes don't often interact with mitochondria, and lysosomes interact only with mitochondria and chloroplasts when they degrade those organelles. Choice (D) is incorrect because the nucleolus doesn't interact with the smooth endoplasmic reticulum, lysosomes, or centrioles.

7. **C** Estrogen is said to be lipid-soluble. This means it can slip through the membrane and bind to an intracellular receptor. A noncompetitive inhibitor would bind to an estrogen receptor, reducing the effectiveness of this binding. Choice (A) is incorrect because estrogen binds to intracellular receptors, not receptors at the plasma membrane. Choice (B) is incorrect because testosterone and estrogen have different effects, but testosterone would not change estrogen effectiveness. Choice (D) is also incorrect because wiping out the ovaries would eliminate, rather than reduce, estrogen levels.

8. **B** You would need a cell that makes and secretes a lot of protein or hormones since the ER moves things out of the cell. Bacteria, (C), do not have organelles. Choice (D), blood cells, is not the best answer because leukocytes do not secrete large amounts of hormones or proteins. Choice (A), neurons, is a potential answer, but (B) is the best answer because the pancreas is a gland that secretes copious amounts of protein and hormones, such as insulin.

9. **A** Choice (A) is correct because contractile vacuoles expel water that accumulates in cells in a hypotonic environment. The strategy described in (B) is wrong because it would increase water intake in *Paramecium*. Choice (C) is true for *Paramecium* but does not help with water gain. Choice (D) would help it find even more dangerously hypotonic environments.

10. **B** ATP is consumed by the Na⁺K⁺ pump, so (B) is correct. The sodium-potassium pump actively pumps both ions. A cotransporter requires one to be passive and one to be active. Choice (A) is not true. ATP is hydrolyzed and not produced. Choice (C) is not true. Choice (D) is incorrect since ATP is hydrolyzed.

CHAPTER 6 DRILL: CELLULAR ENERGETICS

1. **B** The Krebs cycle occurs primarily in the matrix of the mitochondria. The inner membrane and the intermembrane space—(A) and (C), respectively—are used in oxidative phosphorylation.

2. **D** Inhibitor Y is binding at a site outside of the active site and is inducing a conformational change in the enzyme structure. By binding outside of the active site, it must be an allosteric inhibitor, eliminating (A) and (C). Because the inhibitor is binding outside of the active site, it is not competing with the substrate for binding, so it is considered a non-competitive inhibitor.

3. **B** Enzymes are biological catalysts, which lower the activation energy (the energy threshold that must be met to proceed from reactant to product). The reaction coordinate diagram must reflect a decrease in the activation energy, eliminating (C). Furthermore, the enzyme does not alter the energy of the reactants or products, eliminating (A) and (D).

4. **B** Based on the pathway provided, consumption of one glucose and two ATP results in production of four ATP. In other words, each glucose results in a net gain of two ATP. Therefore, two glucose molecules would result in a net gain of four ATP.

5. **B** In fermentation, pyruvic acid is converted into either ethanol or lactic acid. During this process, NADH is recycled into NAD^+.

6. **B** Glycolysis results in the production of ATP (energy), so it is considered an exergonic process.

7. **B** Thermophilic bacteria live in hot environments. Also, a DNA polymerase replicates DNA. Therefore, in the PCR technique, the stage in which DNA is elongated by a DNA polymerase is part three of the cycle, at 72°C. Therefore, bacteria growing in hydrothermal vents between 70–75°C is the answer. The conditions described in (A) and (D) do not reflect the data, which focuses on temperature. Choice (C) describes hot springs, but the temperature does not reflect the enzyme activity most likely for *Taq* polymerase.

8. **D** The free energy does not change between catalyzed and uncatalyzed reactions; therefore, (A) is incorrect. Choice (B) is a correct statement, but it does not answer the question. Choice (C) is incorrect because enzymes catalyze reactions.

9. **B** ATP production will increase in the treated mitochondria because the low pH provides more H^+ ions in the solution. Also, oxygen provides a terminal electron acceptor for oxidative phosphorylation.

10. **D** Choices (A) and (B) are clearly wrong, as they defy physics. Choice (C) is not correct because the anabolic and catabolic reactions are not necessarily equal nor are they direct opposites. Choice (D) is the correct answer because it explains the increase in entropy. Even though organisms build and develop as ordered systems, heat is lost continuously. Additionally, organisms exhale gases and produce waste products that balance the effect of order.

CHAPTER 7 DRILL: MOLECULAR BIOLOGY

1. **B** Okazaki fragments are generated during DNA replication when the DNA polymerase must create short DNA segments due to its requirement for 5′ to 3′ polymerization. Since the newly discovered yeast cell has 3′ to 5′ activity, there would be no lagging strand and likely no Okazaki fragments.

2. **C** Since the gene is much shorter than expected, a stop codon must have been introduced by mutagenesis. This is an example of a nonsense mutation.

3. **D** The order for DNA replication is helicase, RNA primase, DNA polymerase, and ligase.

4. **C** If an mRNA codon is UAC, the complementary segment on a tRNA anticodon is AUG.

5. **D** During post-translational modification, the polypeptide undergoes a conformational change. Choices (A), intron excision, and (B), poly(A) addition, are examples of post-transcriptional modifications. Formation of peptide bonds occurs during translation, not afterward.

6. **B** If 21 nucleotides compose a sequence and 3 nucleotides compose each codon, there would be 7 codons and thus a maximum of 7 amino acids.

7. **A** Choice (B) is incorrect because bacteria make membrane proteins that reside on the plasma membrane. Choice (C) is also incorrect because bacteria use transcription factors. Choice (D) is possible, but (A) is the best answer.

8. **C** Choice (A) is likely incorrect. Choice (B) would be deleterious for the cell. Choice (D) is also incorrect, as DNA replication occurs only during or preceding binary fission.

9. **B** Viruses do not have their own ribosomes, flagellum, or independent metabolism, so (A), (C), and (D) are incorrect. The only possible answer is (B).

10. **C** Transformation occurs when bacteria take up DNA from their surroundings. The pathogenic bacteria did not come alive again, as suggested by (A). Protein was not the transforming agent, so (B) is incorrect. Finally, (D) does not make sense, as genes cannot turn into other genes simply by being in a different cell context.

11. **C** If there are DNA/RNA fragments, then helicase, (A), is present and has opened the strands. RNA primase, (D), is also present because RNA primers have been added. DNA polymerase, (B), must be present because there are small chunks and long chunks. If there was not polymerase, then there wouldn't be small chunks because nothing new would be made. It is likely that DNA ligase (C) is missing and couldn't attach the short Okazaki fragments and the longer leading strand fragments.

CHAPTER 8 DRILL: CELL REPRODUCTION

1. **C** During anaphase, the chromatids are separated by shortening of the spindle fibers. Chemically blocking the shortening of these fibers would arrest the cell in metaphase. The cells are arrested in metaphase as indicated by the alignment of the chromosomes in the center of the cell and their attachment to spindle fibers, eliminating (A) and (D). The chromosomes still seem to be attached to the fibers, so there doesn't appear to be dissociation of the fibers, eliminating (B).

2. **C** The synthesis, or S phase, of the cell cycle represents the step in which the genetic material is duplicated. The only phase labeled in the experiment that represents an increase is phase B. Based on the time scale on the x-axis, this phase lasts approximately 30 minutes.

3. **D** Anaphase represents the cell division stage of the cell cycle and would be the phase that occurs right before the amount of genetic material should decrease. Phase D is the phase right before the genetic material would drop, so (D) is the correct answer.

4. **13** The organism has 13 chromosomes in gametic cells, which are haploid. Therefore, a diploid cell would normally have two copies of each chromosome or 26 making 13 homologous pairs. S phase replicates chromatids, but that won't change the number of pairs.

5. **D** Choices (A) and (B) do not occur, but if they did, they would not impact the gametes. Choice (C) is not likely to occur because the mitotic spindles attach only to kinetochore protein complexes at the centromere. Therefore, (D) is the answer.

6. **B** Both (A) and (B) have sperm or ovum as the first type of cell. Because these are both gametes, they will both have half of the DNA of other types of cells. Choices (C) and (D) can thus be eliminated. Muscles and neurons are both terminally differentiated in G_0 arrest. However, liver and taste buds potentially differ. Although liver cells can divide, taste buds divide at a higher rate, so they are most likely to be in G_2 phase. Therefore, the answer is (B).

7. **A** Nondisjunction results in major changes to the genome, so (B) and (C) can be eliminated. Deletion of an enhancer, (D), could affect the gene expression and then phenotype. However, in (A), translocation may not produce any effect, as the same information exists in the genome.

CHAPTER 9 DRILL: HEREDITY

1. **D** The father and mother are both AB blood type. Since neither parent has a recessive allele, it is impossible for their child to be O blood type.

2. **B** When the phenotype associated with two traits is mixed, this is considered an example of incomplete dominance. In this case, neither red nor blue is dominant, and the resulting progeny exhibited a mixture of the traits (purple).

3. **B** Crossing the pea plant that is heterozygous for both traits (TtGg) with a plant that is recessive for both traits (ttgg) results in the following possible combinations, each of which should occur 25 percent of the time: TtGg (tall and green), Ttgg (tall and yellow), ttGg (short and green), and ttgg (short and yellow). Using the rules of probability, there is a 1/2 likelihood of it being tall and also a 1/2 likelihood of it being yellow. Multiply them together to get 1/4.

4. **C** Because the woman is a carrier, she must have one normal copy of the X chromosome and one diseased copy. Since the boy will receive an X chromosome from his mother, there is a 50 percent chance that he will receive a diseased copy. Because he doesn't have a second X chromosome, he must have the disease if he receives the diseased X chromosome.

5. **D** They both must have a normal copy of the X chromosome. It is possible that the woman may be affected with hemophilia; however, that scenario is extremely unlikely because such a case would require two diseased copies of the X chromosome.

6. **D** Essentially, transmission of hemophilia to a girl born with Turner syndrome would be very similar to the conditions by which a boy would receive the disease. Both boys and girls with Turner syndrome have only one copy of the X chromosome. Therefore, they both would have a 50 percent chance of receiving the diseased copy of the X chromosome.

7. **B** Choice (A) is true, but it doesn't explain why the parents are normal. Choice (B) is a better explanation. Carriers of recessive genetic diseases frequently do not have a recognizable phenotype because one normal allele provides enough functioning protein to avoid the ill effects of the disease allele. Choice (C) is also possible, but (B) is still the better answer. Finally, (D) is not correct because CLN3 is an autosomal gene.

8. **C** Choice (C) describes a testcross that will actually let the breeder know if the male is heterozygous. If any white spotted pups result from that cross, then the male would have contributed a spotted allele. Choice (B) describes a male that is homozygous for the spot allele, so it's incorrect. Choice (A), stop breeding Speckle, is a way to remove spots from the line, but it will not help you identify males with the spot allele. Choice (D) also will not help you determine which black Labrador males have a spot allele.

9. **A** Choice (B) does not apply to this question. Choices (C) and (D) refer to a gene not following Mendel's Law of Dominance. Choice (A)—genes are linked on the same chromosome—describes a normal reason for why certain traits would not follow Mendel's law of independent assortment.

10. **A** Choices (B) and (C) are unlikely to be true and can be eliminated. Choice (D) is true but does not answer the question. Choice (A) is the easiest and most probable explanation.

CHAPTER 10 DRILL: EVOLUTIONARY BIOLOGY

1. **D** Mammals and cephalopods developed similar eye structures independently due to similar selective pressures. This is an example of convergent evolution.

2. **B** The data provided show a transition toward one extreme (black) and away from another (white). This is an example of directional selection.

3. **B** Based on the data, the amount of white-bodied pepper moths decreased between 1802 and 1902, and the amount of black-bodied pepper moths increased during the same period.

4. **B** Longtail moths were included in the experiment as a control to compare the effects that are not associated with color.

5. **B** If the color of ash or soot produced by the Industrial Revolution were white or light gray, this would likely reverse the trend observed, applying additional selection against the black moths.

6. **0.42** The frequency can be calculated as shown below.

$$p + q = 1$$
$$p + (0.3) = 1$$
$$p = 0.7$$
$$2pq = \text{heterozygotes} = 2(0.7)(0.3) = 0.42$$

7. **B** The mole rats live in the same habitat, but over time there has been speciation as the two groups have chosen different plants as their diet. Therefore, (B) is correct. Choice (A) is incorrect because there is not a habitat barrier; the mole rats have access to each other. Choice (C) is incorrect because the mole rats are capable of breeding; they just do not. Choice (D) is incorrect because the mole rats do not interbreed.

8. **D** Choice (A) is incorrect because this is a small population, so genetic drift is likely. Choice (B) is incorrect because this situation does not describe convergent evolution, and the second part of the answer choice does not accurately describe the change of the population. Choice (D) is correct because the number of people that are homozygous recessives will likely become fewer as the red-head allele is mixed with the more dominant hair colors.

9. **D** Choice (D) is the answer because it is the only choice that does not predict evolution occurring on the island, thus supporting a claim for HW equilibrium. Choices (A), (B), and (C) all prevent a Hardy-Weinberg equilibrium and predict evolution. The population is small, mutations are inevitable, and humans usually do not mate randomly. This population will definitely evolve or undergo genetic drift.

10. **A** Choice (A) is correct because the trait is being selected based on female mating preferences. Longer, bigger tails indicate reproductive fitness. Peahens choose males who are healthy and strong enough to grow these big tails and will hopefully produce the best offspring. Choices (B), (C), and (D) are incorrect because they do not describe the female mating preferences.

CHAPTER 11: ANIMAL STRUCTURE AND FUNCTION

1. **B** Lymph nodes harbor large quantities of lymphocytes. During active infections or cancer, mass expansion of lymphocytes causes enlargement and swelling of lymph nodes. This is why doctors often feel the areas around lymph nodes nearest to where a suspected infection is.

2. **C** The placebo strain was included as a control for the experiment. Because no active antibodies should be produced to the placebo strain, the scientists would be able to compare antibodies elicited specifically to the vaccine strain.

3. **B** B-cells (or plasma cells) are responsible for creating antibodies. These are the primary cells associated with lasting memory to vaccinations.

4. **A** The most practical way to increase statistical significance is to increase the sample size, which is (A). More control variables will only ensure that they haven't introduced bias. Repeating a third challenge would evaluate the length of potential protection, not the extent. Using a different statistical test would simply manipulate the data and not provide true significance.

5. **B** Thorns and spines are examples of innate defenses that plants have to deter herbivores. Choice (C), attractive scents, would have the opposite effect. Stomata allow gas exchange and are unrelated to the question. Long stems may be considered a defense only against certain types of threats. Therefore, (B) is the best answer.

6. **A** Choice (B) is incorrect because helper T-cells do not destroy pathogens. Choice (D) is incorrect, as the goal of the immune system is to prevent further infection. Choice (C) is incorrect because only memory B-cells remain, not plasma B-cells. Choice (A) is the correct answer because B-cells secrete antibodies specific to the pathogen in the complement system. This flags the pathogen for destruction.

7. **B** The membrane potential depolarizes at −50 mV so that the voltage-gated sodium channels can open. Note that the sodium channels open and close more quickly than the potassium channels.

8. **C** The blocking of voltage-gated sodium (Na^+) channels by tetrodotoxin would prevent an action potential in cells and therefore the transmission of a neural response to the subsequent neurons.

9. **D** Choices (A) and (C) describe situations that are not directly related to ouabain. Choice (B) is incorrect, as the cell would slowly become more negative without the sodium potassium pump because potassium would leak out of the cell. Choice (D) is correct.

10. **B** Proteins are made primarily in the cell body of neurons. Dendrites receive postsynaptic signals. Axons and axon terminals transmit action potentials in the presynaptic neuron. Choice (B) is correct.

11. **C** In a neuromuscular junction, the neuron synapses with a muscle. The neuron releases acetylcholine as a neurotransmitter to tell the muscle to contract. Thus, acetylcholinesterase removes extra acetylcholine from the synapse after the transmission. Without this enzyme, the neurotransmitter will build up in the synapse, causing overstimulation of the acetylcholine receptors on the muscle and hyper-contraction of the muscles. Choices (A), (B), and (D) are incorrect because they describe the opposite of what would occur.

12. **A** By definition, the threshold is the point at which enough voltage-gated channels have opened to permit a switch in the polarity of the membrane and an action potential to perpetuate through the axon of the neuron. Choices (B), the release of neurotransmitters, and (C), peak depolarization, both occur after the threshold is met. Choice (D), sodium ions enter the cell during depolarization, is vague and not as strong as (A).

13. **B** Both Schwann cells and oligodendrocytes act to wrap around the axons of neurons to insulate the membrane depolarization and expedite the action potential. Choice (B) is correct.

14. **B**   Choices (A), (C), and (D) are incorrect because testosterone and estradiol will easily diffuse across the membrane, as they are hydrophobic, steroid-type molecules. Receptors are not needed on the outside of the plasma membrane. Instead, receptors inside the cell will receive and transduce the signal. Choice (B) is the correct answer.

15. **B**   Choice (B) is the correct answer because the receptor for vasopressin is in the G-protein–coupled receptor family. These receptors sit on the plasma membrane. Only cells with the vasopressin receptor will respond to the presence of vasopressin, even if other cells are exposed to the hormone. Therefore, (A) and (C) are incorrect. Choice (D) is also incorrect because vasopressin does not enter cells.

CHAPTER 12 DRILL: BEHAVIOR AND ECOLOGY

1. **D**   The behavior displayed by the chimpanzee represents insight because the chimpanzee has figured out how to solve the problem without external influence or learning.

2. **A**   Viruses would display reproductive strategies most similar to r-strategists because they aim to reproduce as fast as possible and create as many progeny as possible in order to increase their odds of transmission to other hosts.

3. **C**   They would be considered secondary consumers because they consume the primary consumers (plankton), which consume algae (the producers).

4. **C**   Since brook trout can tolerate pH values as low as 4.9 and do not appear to diminish, the pH of the creek must exceed 4.9, so we can eliminate (D). Since crayfish cannot tolerate pH levels lower than 5.4, the pH of the stream must have dropped lower than this value. Only (C) falls in this range.

5. **C**   Only acid rain would directly explain why the pH would drop in the creek over the time period.

6. **D**   Based on Table 1, the pH must exceed 6.1 for snails to be able to return to the creek ecosystem.

7. **D**   Choices (A) and (B) refer to light and gravity responses of plants. Choice (C) is not an actual tropism—it is a made-up word. Thigmotropism is the term for how plants respond to touch. The correct answer is (D).

8. **D** If (D) is a correct statement, then simply cutting out beef and lamb from your diet can dramatically decrease your carbon footprint—even if you do not decide to eat vegan. Choices (A), (B), and (C) are all true statements with respect to reducing the carbon footprint of food.

9. **B** A community that has been destroyed and then rebuilds is a result of secondary succession. Choice (A) would be the answer only if there were no life present before the new growth. Choice (C) is true only of a mature forest. Choice (D) is incorrect, as the species taking over the field is native to the rainforests on the periphery of the field. Nothing in the question suggests invasive species. Therefore, (B) is correct.

10. **A** If the intestines are occupied with beneficial bacteria, there will not be any room available for pathogenic bacteria to grow by density-dependent limitations on the population. Choice (B) is not accurate because pathogenic bacteria do not require a niche prepared for them by probiotic bacteria. Choice (C) is true, but it does not answer the question. Choice (D) is also incorrect, leaving (A) as your answer.

CHAPTER 13 DRILL: QUANTITATIVE SKILLS AND BIOSTATISTICS

1. **A** In order to answer this question, you must first calculate how much weight each subject lost and then divide by the number of subjects (in this case, five).

Subject	Starting Weight (pounds)	Final Weight (pounds)	Weight Lost (pounds)
1	184	176	8
2	200	190	10
3	221	225	–4
4	235	208	27
5	244	225	19

Note that subject 3 *gained* four pounds. Total weight lost is 60 pounds (remember to subtract 4 pounds for subject 3, not add), divided by 5 subjects is 12 pounds. The average weight lost is 12 pounds ((A) is correct).

2. **C** Plant 6 is 68 inches. All the answer choices list median values smaller than this, so the answer must start with "Yes" (eliminate (B) and (D)). In order to determine the median height for all six plants, their heights must first be organized in ascending order: 61, 66, 67, 68, 70, 72. The middle two numbers are 67 and 68; when averaged, this produces a median of 67.5 inches ((C) is correct). Notice that if all the data points are whole numbers, the median value must either end in .0 or .5, so you can quickly eliminate (A) and (B) in this question. In fact, (A) and (B) are giving the value for mean, not median.

3. **D** The test scores ordered from smallest to largest are:

32, 65, 66, 67, 68, 68, 69, 70, 71, 72, 73, 75, 75, 75, 78, 82

The most frequently recurring number in the set above is 75, so this is the mode (eliminate (A) and (B)). The smallest number is 32 and the largest is 82. The range is the difference between the two, or 50 ((D) is correct).

4. **D** This question is testing the product rule.

$$P(AABbCC) = P(AA) \times P(Bb) \times P(CC)$$
$$P = \frac{1}{4} \times \frac{1}{2} \times \frac{1}{4}$$
$$P = \frac{1}{32}$$

5. **A** This question is testing the sum rule, but you also need to use the product rule.

$$P(Aabb \text{ or } aaBb) = P(Aabb) + P(aaBb)$$

$$P = [P(Aa) \times P(bb)] + [P(aa) \times P(Bb)]$$
$$P = \left(\frac{1}{2}\right)\left(\frac{1}{2}\right) + \left(\frac{1}{2}\right)\left(\frac{1}{2}\right)$$
$$= \frac{1}{4} + \frac{1}{4}$$
$$P = \frac{2}{4} = \frac{1}{2}$$

6. **C** In chi-square tests, a calculated x^2 value is compared to a critical value (CV) from a chi-square table, like the one on the AP Biology Equations and Formulas sheet. If $x^2 < CV$, you accept the null hypothesis (H_o). If $x^2 > CV$, you reject H_o. Our H_o is that the observed and expected data match and that the experimental plants have a 3:1 ratio of yellow to green plants. Based on this, the only possible answer choices are (B) and (C); the other choices mix up how x^2 values and critical values are compared (eliminate (A) and (D)).

One hundred plants were studied, and we expect three-fourths of them to be yellow (75 plants) and one-fourth of them to be green (25 plants). Next, we compare expected (E) and observed (O) data and calculate a x^2 value.

Plant Phenotype	Observed (O)	Expected (E)	$(O - E)^2/E$
Yellow	84	75	1.08
Green	16	25	3.24
Total	**100**	**100**	**4.32**

Because two possibilities are being compared (green and yellow), the degrees of freedom in this test = 2 − 1 = 1. Using $p = 0.05$, the critical value is 3.84 (which you don't need to look up because it is listed in all answer choices). Since $x^2 >$ CV, you reject H_o. The observed data do not match the expected data, and the correct answer is (C).

7. **B** The first thing you need to do is sort out the alleles for each gene. Since the F1 generation had long phenotypes, you know these must be dominant to the short phenotypes. Let's use:

L = long leaves
l = short leaves
S = long shoots
s = stubby shoots

The parental cross must have been $LLSS \times llss$, and all F1 plants were $LlSs$. Our null hypothesis (H_o) is that the two genes are segregating independently. This means the expected ratio of F2 phenotypes will be $9LS:3Ls:3lS:1ls$. Since 1,000 F2 plants were generated, we expect:

$P(\text{long leaves, long shoots}) = P(LS \text{ phenotype}) = \dfrac{9}{16} \times 1{,}000 = 562.5$

$P(\text{long leaves, stubby shoots}) = P(Ls \text{ phenotype}) = \dfrac{3}{16} \times 1{,}000 = 187.5$

$P(\text{short leaves, long shoots}) = P(lS \text{ phenotype}) = \dfrac{3}{16} \times 1{,}000 = 187.5$

$P(\text{short leaves, stubby shoots}) = P(ls \text{ phenotype}) = \dfrac{1}{16} \times 1{,}000 = 62.5$

Next, generate a chart:

F2 Phenotype	F2 Genotype	Observed (O)	Expected (E)	$(O - E)^2/E$
Long leaves, long shoots	$L{-}S{-}$	382	562.5	57.9
Long leaves, stubby shoots	$L{-}ss$	109	187.5	32.9
Short leaves, long shoots	$llS{-}$	112	187.5	30.4
Short leaves, stubby shoots	$llss$	397	62.5	1,790.2
Total		1,000	1,000	1,911.4

$x^2 = 1{,}911.4$ and degrees of freedom = # of possibilities − 1 = 4 − 1 = 3 (eliminate (C) and (D)). Using $p = 0.05$, we get a critical value of 7.82. Since $x^2 >$ CV, we reject H_o. These two genes are not segregating independently (eliminate (A); (B) is correct).

AP BIOLOGY
EQUATIONS AND FORMULAS

Statistical Analysis and Probability

Mean

$$\bar{x} = \frac{1}{n} \sum_{i=1}^{n} x_i$$

Standard Deviation*

$$S = \sqrt{\frac{\sum (x_i - \bar{x})^2}{n-1}}$$

Standard Error of the Mean*

$$SE_{\bar{x}} = \frac{S}{\sqrt{n}}$$

Chi-Square

$$x^2 = \sum \frac{(o-e)^2}{e}$$

Chi-Square Table

p-value	Degrees of Freedom							
	1	2	3	4	5	6	7	8
0.05	3.84	5.99	7.82	9.49	11.07	12.59	14.07	15.51
0.01	6.64	9.21	11.34	13.28	15.09	16.81	18.48	20.09

$\bar{x}$ = sample mean

n = size of the sample

s = sample standard deviation (i.e., the sample-based estimate of the standard deviation of the population)

o = observed results

e = expected results

Degrees of freedom are equal to the number of distinct possible outcomes minus one.

Laws of Probability
If A and B are mutually exclusive, then:
$$P(A \text{ or } B) = P(A) + P(B)$$
If A and B are independent, then:
$$P(A \text{ and } B) = P(A) \times P(B)$$

Hardy-Weinberg Equations

$p^2 + 2pq + q^2 = 1$ p = frequency of the dominant allele in a population

$p + q = 1$ q = frequency of the recessive allele in a population

Metric Prefixes

Factor	Prefix	Symbol
10^9	giga	G
10^6	mega	M
10^3	kilo	k
10^{-2}	centi	c
10^{-3}	milli	m
10^{-6}	micro	μ
10^{-9}	nano	n
10^{-12}	pico	p

Mode = value that occurs most frequently in a data set

Median = middle value that separates the greater and lesser halves of a data set

Mean = sum of all data points divided by the number of data points

Range = value obtained by subtracting the smallest observation (sample minimum) from the greatest (sample maximum)

* For the purposes of the AP Exam, students will not be required to perform calculations using this equation; however, they must understand the underlying concepts and applications.

Rate and Growth

Rate

$$\frac{dY}{dt}$$

Population Growth

$$\frac{dN}{dt} = B - D$$

Exponential Growth

$$\frac{dN}{dt} = r_{max}N$$

Logistic Growth

$$\frac{dN}{dt} = {}^r max^N \left(\frac{K - N}{K}\right)$$

Temperature Coefficient Q_{10} †

$$Q_{10} = \left(\frac{k_2}{k_1}\right)^{\frac{10}{T_2 - T_1}}$$

Primary Productivity Calculation

$$\frac{mg\ O_2}{L} \times \frac{0.698\ mL}{mg} = \frac{mL\ O_2}{L}$$

$$\frac{mL\ O_2}{L} \times \frac{0.536\ mg\ C\ fixed}{mL\ O_2} = \frac{mg\ C\ fixed}{L}$$

(at standard temperature and pressure)

---

dY = amount of change

dt = change in time

B = birth rate

D = death rate

N = population size

K = carrying capacity

r_{max} = maximum per capita growth rate of population

T_2 = higher temperature

T_1 = lower temperature

k_2 = reaction rate at T_2

k_1 = reaction rate at T_1

Q_{10} = the factor by which the reaction rate increases when the temperature is raised by ten degrees

---

Water Potential (ψ)

$$\psi = \psi_P + \psi_S$$

ψ_P = pressure potential

ψ_S = solute potential

The water potential will be equal to the solute potential of a solution in an open container because the pressure potential of the solution in an open container is zero.

The Solute Potential of a Solution

$$\psi_S = -iCRT$$

i = ionization constant (This is 1.0 for sucrose because sucrose does not ionize in water.)

C = molar concentration

R = pressure constant ($R = 0.0831$ liter bars/mole K)

T = temperature in Kelvin (C + 273)

Surface Area and Volume

Volume of a Sphere

$$V = \frac{4}{3}\pi r^3$$

Volume of a Rectangular Solid

$$V = lwh$$

Volume of a Right Cylinder

$$V = \pi r^2 h$$

Surface Area of a Sphere

$$A = 4\pi r^2$$

Surface Area of a Cube

$$A = 6a^2$$

Surface Area of a Rectangular Solid

$A = \sum$ surface area of each side

---

r = radius

l = length

h = height

w = width

s = length of one side of a cube

A = surface area

V = volume

$\sum$ = sum of all

Dilution (used to create a dilute solution from a concentrated stock solution)

$$C_i V_i = C_f V_f$$

i = initial (starting) C = concentration of solute

f = final (desired) V = volume of solution

Gibbs Free Energy

$$\Delta G = \Delta H - T\Delta S$$

ΔG = change in Gibbs free energy

ΔS = change in entropy

ΔH = change in enthalpy

T = absolute temperature (in Kelvin)

$$pH^* = -\log_{10}[H^+]$$

---

* For the purposes of the AP Exam, students will not be required to perform calculations using this equation; however, they must understand the underlying concepts and applications.

† For use with labs only (optional)

Part VI
Practice Test 2

Practice Test 2
Practice Test 2: Answers and Explanations

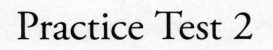

Practice Test 2

Completely darken bubbles with a No. 2 pencil. If you make a mistake, be sure to erase mark completely. Erase all stray marks.

1. YOUR NAME:
(Print)

| Last | First | M.I. |

SIGNATURE: _____ DATE: ___/___/___

HOME ADDRESS: _____
(Print)

Number and Street

_____
City   State   Zip Code

PHONE NO. : _____
(Print)

5. YOUR NAME

| First 4 letters of last name | | | | FIRST INIT | MID INIT |

(Bubbles A–Z for each column)

IMPORTANT: Please fill in these boxes exactly as shown on the back cover of your test book.

2. TEST FORM

3. TEST CODE   **4. REGISTRATION NUMBER**

(Number bubbles 0–9; Test Code letter column A–G)

6. DATE OF BIRTH

Month	Day	Year
JAN		
FEB		
MAR		
APR		
MAY		
JUN		
JUL		
AUG		
SEP		
OCT		
NOV		
DEC		

7. SEX
MALE
FEMALE

The Princeton Review®

1 A B C D
2 A B C D
3 A B C D
4 A B C D
5 A B C D
6 A B C D
7 A B C D
8 A B C D
9 A B C D
10 A B C D
11 A B C D
12 A B C D
13 A B C D
14 A B C D
15 A B C D
16 A B C D

17 A B C D
18 A B C D
19 A B C D
20 A B C D
21 A B C D
22 A B C D
23 A B C D
24 A B C D
25 A B C D
26 A B C D
27 A B C D
28 A B C D
29 A B C D
30 A B C D
31 A B C D
32 A B C D

33 A B C D
34 A B C D
35 A B C D
36 A B C D
37 A B C D
38 A B C D
39 A B C D
40 A B C D
41 A B C D
42 A B C D
43 A B C D
44 A B C D
45 A B C D
46 A B C D
47 A B C D
48 A B C D

49 A B C D
50 A B C D
51 A B C D
52 A B C D
53 A B C D
54 A B C D
55 A B C D
56 A B C D
57 A B C D
58 A B C D
59 A B C D
60 A B C D
61 A B C D
62 A B C D
63 A B C D

64   65   66   67   68   69

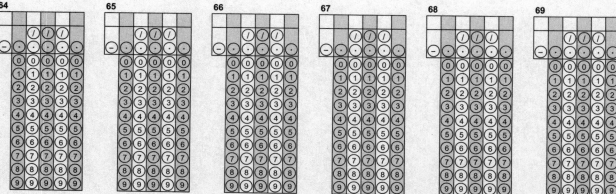

AP® Biology Exam

SECTION I: Multiple-Choice Questions

DO NOT OPEN THIS BOOKLET UNTIL YOU ARE TOLD TO DO SO.

At a Glance

Total Time
1 hour and 30 minutes
Number of Questions
69
Percent of Total Score
50%
Writing Instrument
Pencil required

Instructions

Section I of this examination contains 69 multiple-choice questions. These are broken down into Part A (63 multiple-choice questions) and Part B (6 grid-in questions).

Indicate all of your answers to the multiple-choice questions on the answer sheet. No credit will be given for anything written in this exam booklet, but you may use the booklet for notes or scratch work. After you have decided which of the suggested answers is best, completely fill in the corresponding oval on the answer sheet. Give only one answer to each question. If you change an answer, be sure that the previous mark is erased completely. Here is a sample question and answer.

Sample Question   Sample Answer

Chicago is a   Ⓐ ● Ⓒ Ⓓ

(A)   state
(B)   city
(C)   country
(D)   continent

Use your time effectively, working as quickly as you can without losing accuracy. Do not spend too much time on any one question. Go on to other questions and come back to the ones you have not answered if you have time. It is not expected that everyone will know the answers to all the multiple-choice questions.

About Guessing

Many candidates wonder whether or not to guess the answers to questions about which they are not certain. Multiple-choice scores are based on the number of questions answered correctly. Points are not deducted for incorrect answers, and no points are awarded for unanswered questions. Because points are not deducted for incorrect answers, you are encouraged to answer all multiple-choice questions. On any questions you do not know the answer to, you should eliminate as many choices as you can, and then select the best answer among the remaining choices.

BIOLOGY
SECTION I
69 Questions
Time—90 minutes

Directions: Each of the questions or incomplete statements below is followed by four suggested answers or completions. Select the one that is best in each case and then fill in the corresponding oval on the answer sheet.

1. In general, animal cells differ from plant cells in that only animal cells

    (A) perform cellular respiration.
    (B) contain transcription factors.
    (C) do not contain vacuoles.
    (D) lyse when placed in a hypotonic solution.

2. A cell from the leaf of the aquatic plant Elodea was soaked in a 15 percent sugar solution, and its contents soon separated from the cell wall and formed a mass in the center of the cell. All of the following statements are true about this event EXCEPT

    (A) the vacuole lost water and became smaller
    (B) the space between the cell wall and the cell membrane expanded
    (C) the large vacuole contained a solution with much lower water potential than that of the sugar solution
    (D) the concentration of solutes in the extracellular environment is hypertonic with respect to the cell's interior

3. A chemical agent is found to denature acetylcholinesterase in the synaptic cleft. What effect will this agent have on the neurotransmitter, acetylcholine?

    (A) Acetylcholine will not be released from the presynaptic membrane.
    (B) Acetylcholine will not bind to receptor proteins on the postsynaptic membrane.
    (C) Acetylcholine will not diffuse across the cleft to the postsynaptic membrane.
    (D) Acetylcholine will not be degraded in the synaptic cleft.

4. The base composition of DNA varies from one species to another. Which of the following ratios would you expect to remain constant in the DNA?

    (A) Cytosine : Adenine
    (B) Pyrimidine : Purine
    (C) Adenine : Guanine
    (D) Guanine : Deoxyribose

Questions 5–6 refer to the following passage.

Consider the following pathway of reactions catalyzed by enzymes (shown in numbers):

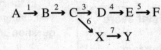

5. Which of the following situations represents feedback inhibition?

    (A) Protein D activating enzyme 4
    (B) Protein B stimulating enzyme 1
    (C) Protein 7 inhibiting enzyme C
    (D) Protein X inhibiting enzyme 2

6. An increase in substance F leads to the inhibition of enzyme 3. All of the following are direct or indirect results of the process EXCEPT

    (A) an increase in substance X
    (B) increased activity of enzyme 6
    (C) decreased activity of enzyme 4
    (D) increased activity of enzyme 5

7. If a competitive inhibitor is bound to enzyme 1, which of the following would be decreased?

    I. Protein A
    II. Protein C
    III. Protein X

    (A) I only
    (B) II only
    (C) II and III
    (D) I, II, and III

GO ON TO THE NEXT PAGE.

Questions 8–11 refer to the following passage.

The affinity of hemoglobin for oxygen is reduced by many factors, including low pH and high CO_2. The graph below shows the different dissociation curves that maternal (normal) hemoglobin and fetal hemoglobin have.

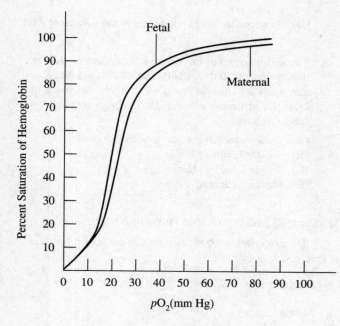

8. Based on the graph, it can be concluded that

(A) fetal hemoglobin surrenders O_2 more readily than maternal hemoglobin

(B) the dissociation curve of fetal hemoglobin is to the right of maternal hemoglobin

(C) fetal hemoglobin has a higher affinity for O_2 than does maternal hemoglobin

(D) fetal and maternal hemoglobin differ in structure

9. Which of the following processes would likely shift the normal dissociation curve to the right?

(A) Photosynthesis

(B) Respiration

(C) Fermentation

(D) Mitosis

10. Hemoglobin's affinity for O_2

(A) decreases as blood pH decreases

(B) increases as H^+ concentration increases

(C) increases as blood pH decreases

(D) decreases as OH^- concentration increases

11. How much pO_2 would it take in an extremely CO_2-rich environment to saturate hemoglobin 90 percent?

(A) 15

(B) 30

(C) 45

(D) 60

12. All of the following are differences between prokaryotes and eukaryotes EXCEPT

(A) eukaryotes have linear chromosomes, while prokaryotes have circular chromosomes

(B) eukaryotes possess double-stranded DNA, while prokaryotes possess single-stranded DNA

(C) eukaryotes process their mRNA, while in prokaryotes, transcription and translation occur simultaneously

(D) eukaryotes contain membrane-bound organelles, while prokaryotes do not

13. In minks, the gene for brown fur (B) is dominant over the gene for silver fur (b). Which set of genotypes represents a cross that could produce offspring with silver fur from parents that both have brown fur?

(A) BB × BB

(B) BB × Bb

(C) Bb × Bb

(D) Bb × bb

14. All viruses contain at least these two principal components:

(A) DNA and proteins

(B) nucleic acids and a capsid

(C) DNA and a cell membrane

(D) RNA and a cell wall

15. All of the following are examples of hydrolysis EXCEPT

(A) conversion of fats to fatty acids and glycerol

(B) conversion of proteins to amino acids

(C) conversion of starch to simple sugars

(D) conversion of pyruvic acid to acetyl-CoA

16. In cells, which of the following can catalyze reactions involving hydrogen peroxide, provide cellular energy, and make proteins, in that order?

(A) Peroxisomes, mitochondria, and ribosomes

(B) Peroxisomes, mitochondria, and lysosomes

(C) Peroxisomes, mitochondria, and Golgi apparatus

(D) Lysosomes, chloroplasts, and ribosomes

GO ON TO THE NEXT PAGE.

Questions 17 and 18 refer to the following passage.

The Loop of Henle is a structure within each of the million nephrons within a kidney. As shown in the figure, the two sides have different permeabilities, and there is differential movement across each membrane. The Loop acts as a counter-current multiplier that makes the medulla of the kidney very osmotic. The longer the loop, the higher and more powerful the osmolarity gradient that is created. The gradient is required for the reclamation of water from the urine collecting duct. On the right side of the figure is the urine collecting duct. If the body needs to retain water, anti-diuretic hormone makes this region permeable to water via the introduction of aquaporins, and the osmotic pull of the medulla reclaims the water, out of the collecting duct, which makes the urine more concentrated.

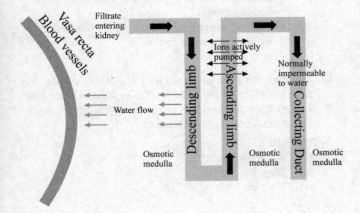

17. Which of the following statements correctly describes the state of things near the top of the descending limb?

(A) The fluid within the descending limb is hypotonic to fluid in the space surrounding the tubule.

(B) The blood within the vasa recta is hypotonic to the filtrate within the descending limb.

(C) The fluid in the area surrounding the tubule is hypertonic to the blood in the vasa recta.

(D) The water in the area surrounding the tubule has a higher water potential than the water in the descending tubule.

18. What type of transport is occurring when water flows out of the descending tubule?

(A) Simple diffusion
(B) Facilitated diffusion
(C) Active transport
(D) Secondary active transport

19. If an inhibitor of anti-diuretic hormone, such as caffeine, was ingested, what would be the result?

(A) Aquaporins would appear in the collecting duct.
(B) Aquaporins would increase in number in the collecting duct.
(C) It would block aquaporins like a competitive inhibitor.
(D) Aquaporins would not appear in the collecting duct.

20. The ability to reclaim water from the collecting duct is directly related to the osmotic pull of the medulla. Kangaroo rats are know to produce extremely concentrated urine. Compared to a human, the nephrons in kangaroo rats must have

(A) thick walls that are impermeable to water
(B) shorter Loops of Henle
(C) longer Loops of Henle
(D) shorter collecting ducts

Questions 21 and 22 refer to the following graph.

The graph below shows two growth curves for bacterial cultures, A and B.

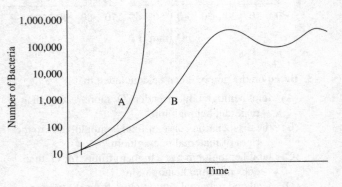

21. Which of the following represents the carrying capacity for culture B?

(A) 10
(B) 50
(C) 100,000
(D) 1,000,000

22. What could explain the differnce between culture A and culture B?

(A) Culture B started with more bacteria than culture A.
(B) Culture A was grown with a competitive inhibitor.
(C) Culture B was not measured as often.
(D) Culture A has not yet exhausted its space and resources.

GO ON TO THE NEXT PAGE.

23. In plants, the process of phototropism would be best exemplified by

    (A) a sunflower photographed pointing one direction in the morning and another direction at night.
    (B) a vine climbing and twisting around a trellis.
    (C) a flowering tree that blooms only in the springtime.
    (D) a flower that changes color to attract more pollinators.

24. Females with Turner's syndrome have a high incidence of hemophilia, a recessive, X-linked trait. Based on this information, it can be inferred that females with this condition

    (A) have an extra X chromosome
    (B) have an extra Y chromosome
    (C) have one less X chromosome than normal
    (D) have one less Y chromosome than normal

25. When a retrovirus inserted its DNA into the middle of a bacterial gene, it altered the normal reading frame by one base pair. This type of mutation is called

    (A) duplication
    (B) translocation
    (C) inversion
    (D) frameshift mutation

26. The principal inorganic compound found in living things is

    (A) carbon
    (B) oxygen
    (C) water
    (D) glucose

27. Metafemale syndrome, a disorder in which a female has an extra X chromosome, is the result of nondisjunction. This failure in oogenesis would first be apparent when

    (A) the homologous chromosomes are lined up in the middle
    (B) the sister chromatids are lined up in the middle
    (C) the nuclear envelope breaks down before meiosis
    (D) the homologous chromosomes are pulling apart

28. All of the following are modes of asexual reproduction EXCEPT

    (A) sporulation
    (B) fission
    (C) budding
    (D) meiosis

29. Invertebrate immune systems possess which of the following?

    (A) Killer T-cells
    (B) Phagocytes
    (C) B-cells
    (D) Helper T-cells

Questions 30–32 refer to the following passage.

The following are important pieces of replication and transcription machinery:

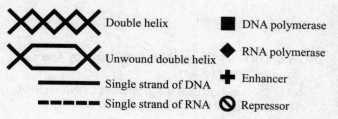

Double helix
Unwound double helix
Single strand of DNA
Single strand of RNA
DNA polymerase
RNA polymerase
Enhancer
Repressor

30. Which of the following figures would be present without helicase?

    (A)
    (B)
    (C)
    (D)

31. What situation might have occurred to produce the following situation?

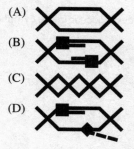

    (A) Stalled DNA replication
    (B) Initiation of transcription
    (C) Repression of transcription
    (D) Crossing-over in meiosis

32. Put these four situations in the correct order:

    I.
    II.
    III.
    IV.

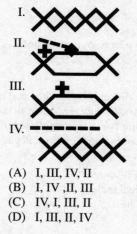

    (A) I, III, IV, II
    (B) I, IV, II, III
    (C) IV, I, III, II
    (D) I, III, II, IV

GO ON TO THE NEXT PAGE.

33. Photosynthesis requires

    (A) glucose, light, CO_2
    (B) light, CO_2, water
    (C) water, soil, O_2
    (D) O_2, water, light

34. Which of following conclusions was made by Avery, MacLeod, and McCarty?

    (A) The double helix is the structure of a DNA molecule.
    (B) DNA is the hereditary material.
    (C) DNA replication is semi-conservative.
    (D) DNA polymerase adds bases to only the 3' side of DNA.

35. Which of the following processes occur in the cytoplasm of an eukaryotic cell?

    I. DNA replication

    II. Transcription

    III. Translation

    (A) I only
    (B) III only
    (C) II and III only
    (D) I, II, and III

36. Crossing-over during meiosis permits scientists to determine

    (A) the chance for variation in zygotes
    (B) the rate of mutations
    (C) the distance between genes on a chromosome
    (D) which traits are dominant or recessive

37. An animal cell that is permeable to water but not salts has an internal NaCl concentration of 10%. If placed in freshwater the cell will

    (A) plasmolyze
    (B) swell and eventually lyse
    (C) endocytose water into a large central vacuole
    (D) shrivel

38. Three distinct bird species, flicker, woodpecker, and elf owl, all inhabit a large cactus, *Cereus giganteus*, in the desert of Arizona. Since competition among these birds rarely occurs, the most likely explanation for this phenomenon is that these birds

    (A) have a short supply of resources
    (B) have different ecological niches
    (C) do not live together long
    (D) are unable to breed

39. Lampreys attach to the skin of lake trout and absorb nutrients from its body. This relationship is an example of

    (A) commensalism
    (B) parasitism
    (C) mutualism
    (D) gravitropism

40. The nucleotide sequence of a template DNA molecule is 5'-C-A-T-3'. A mRNA molecule with a complementary codon is transcribed from the DNA. What would the sequence of the anticodon that binds to this mRNA be?

    (A) 5'-G-T-A-3'
    (B) 5'-G-U-A-3'
    (C) 5'-C-A-U-3'
    (D) 5-'U-A-C-3'

41. Viruses are considered an exception to the cell theory because they

    (A) are not independent organisms
    (B) have only a few genes
    (C) move about via their tails
    (D) have evolved from ancestral protists

GO ON TO THE NEXT PAGE.

Questions 42–44 refer to the following passage.

It is difficult to determine exactly how life began. Answer the following questions as if the following data had been collected billions of years ago.

	4 billion years ago	3.5 billion years ago	3.25 billion years ago	3 billion years ago
Atmospheric carbon dioxide	Present	Present	Present	Present
Atmospheric oxygen	Absent	Absent	Absent	Present
RNA	Absent	Present	Present	Present
Protein	Absent	Absent	Present	Present
DNA	Absent	Absent	Present	Present
Life	Absent	Present	Present	Present

42. Which best describes the origin of life according to the data?

(A) Life required an environment with atmospheric oxygen and any type of nucleic acids.
(B) Life required an environment with atmospheric carbon dioxide, but not atmospheric oxygen.
(C) Life required an environment with self-replicating nucleic acids that can take on many shapes.
(D) Life required an environment with nucleic acids and proteins, but not atmospheric oxygen.

43. When is the earliest that functional ribosomes could have been found?

(A) Between 4 billion and 3.5 billion years ago
(B) Between 3.5 billion and 3.25 billion years ago
(C) Between 3.25 billion and 3 billion years ago
(D) Between 3 billion years ago and the present

44. Photosynthesis likely began_____ billion years ago when the first _____ appeared.

(A) 4.5; autotrophs
(B) 3.2; autotrophs
(C) 3.5; heterotrophs
(D) 3.2; heterotroph

45. The sequence of amino acids in hemoglobin molecules of humans is more similar to that of chimpanzees than it is to the hemoglobin of dogs. This similarity suggests that

(A) humans and dogs are more closely related than humans and chimpanzees
(B) humans and chimpanzees are more closely related than humans and dogs
(C) humans are related to chimpanzees but not to dogs
(D) humans and chimpanzees are closely analogous

46. Two individuals, one with type B blood and one with type AB blood, have a child. The probability that the child has type O blood is

(A) 0%
(B) 25%
(C) 50%
(D) 100%

Questions 47 and 48 refer to the following bar graph, which shows the relative biomass of four different populations of a particular food pyramid.

Relative Biomass

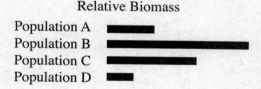

Population A
Population B
Population C
Population D

47. The largest amount of energy is available to

(A) population A
(B) population B
(C) population C
(D) population D

48. Which of the following would be the most likely result if there was an increase in the number of organisms in population C?

(A) The biomass of population D will remain the same.
(B) The biomass of population B will decrease.
(C) The biomass of population A will steadily decrease.
(D) The food source available to population C would increase.

GO ON TO THE NEXT PAGE.

Questions 49–52 refer to the following illustration and information.

The cell cycle is a series of events in the life of a dividing eukaryotic cell. It consists of four stages: G_1, S, G_2, and M. The duration of the cell cycle varies from one species to another and from one cell type to another. The G_1 phase varies the most. For example, embryonic cells can pass through the G_1 phase so quickly that it hardly exists, whereas neurons are arrested in the cell cycle and do not divide.

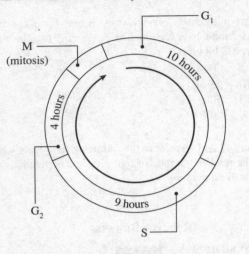

49. During which phase do chromosomes replicate?

(A) G_1
(B) S
(C) G_2
(D) M

50. In mammalian cells, the first sign of mitosis beginning is the

(A) appearance of chromosomes
(B) separation of chromatids
(C) disappearance of the cell membrane
(D) replication of chromosomes

51. If the cell cycle fails to progress, which of the following is NOT a possible explanation?

(A) There are inadequate phosphate groups available for the cyclin dependent kinase.
(B) A tumor suppressor gene has signaled for apoptosis.
(C) A cyclin is unable to release from its cyclin dependent kinase.
(D) An inhibitor of a cyclin gene has been highly expressed.

52. Since neurons are destined never to divide again, what conclusion can be made?

(A) These cells will go through cell division.
(B) These cells will be permanently arrested in the G_1 phase.
(C) These cells will be permanently arrested in the M phase.
(D) These cells will quickly enter the S phase.

GO ON TO THE NEXT PAGE.

Questions 53–56 refer to the following graphs, which show the permeability of (P) ions during an action potential in a ventricular contractile cardiac fiber. The left graph shows the membrane potential changes, and the right shows the corresponding ion permeabilities over the time frame of the action potential.

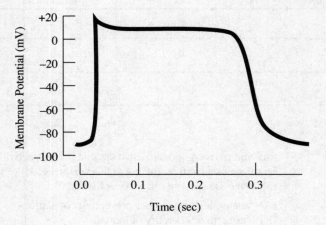
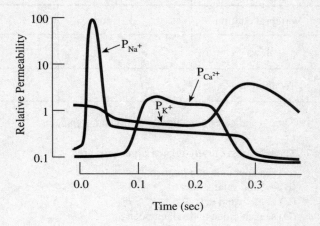

53. Based on the graph, the resting membrane potential of the muscle fibers is closest to

    (A) −90 mV
    (B) −70 mV
    (C) 0 mV
    (D) +70 mV

54. Which of the following statements is true concerning the initial phase of depolarization?

    (A) Voltage-gated K^+ channels open in the plasma membrane.
    (B) The concentration of Ca^{2+} ions within the plasma membrane becomes more negative.
    (C) The membrane potential stays close to −40 mV.
    (D) The permeability to Na^+ ions increases.

55. In cardiac fibers, the duration of an action potential is approximately

    (A) 0.10 secs
    (B) 0.20 secs
    (C) 0.25 secs
    (D) 0.30 secs

56. The action potential of skeletal muscle cells is much shorter than that of cardiac cells. Which of the following facts about cardiac muscle best explains this?

    (A) The membrane is permeable to Na^+, not K^+.
    (B) Voltage-gated K^+ channels open during depolarization, not repolarization.
    (C) Depolarization is prolonged compared to that in skeletal muscle fibers.
    (D) The refractory period is shorter than that of skeletal muscle fibers.

GO ON TO THE NEXT PAGE.

Questions 57 and 58 refer to the data below concerning the general animal body plan of four organisms.

Characteristic	Sea anemone	Hagfish	Eel	Salamander
Vertebral column		+	+	+
Jaws			+	+
Walking legs				+

Note: + indicates a feature present in an organism.

57. The two most closely related organisms are

    (A) sea anemone and hagfish
    (B) eel and salamander
    (C) hagfish and eel
    (D) sea anemone and salamander

58. The correct order of evolution for the traits above is

    (A) jaws – vertebral column – walking legs
    (B) walking legs – jaws – vertebral column
    (C) jaws – walking legs – vertebral column
    (D) vertebral column – jaws – walking legs

59. Pre- and post-zygotic barriers exist that prevent two different species from producing viable offspring. All of the following are pre-zygotic barriers EXCEPT

    (A) anatomical differences preventing copulation
    (B) different temporality of mating
    (C) sterility of offspring
    (D) incompatible mating songs

60. Birds and insects have both adapted wings to travel by flight. The wings of birds and insects are an example of

    (A) divergent evolution
    (B) convergent evolution
    (C) speciation
    (D) genetic drift

GO ON TO THE NEXT PAGE.

Questions 61–63 refer to the following synthetic pathway of nRNA pyrimidine, cytidine 5′ triphosphate, CTP. This pathway begins with the condensation of two small molecules by the enzyme aspartate transcarbamylase (ATCase).

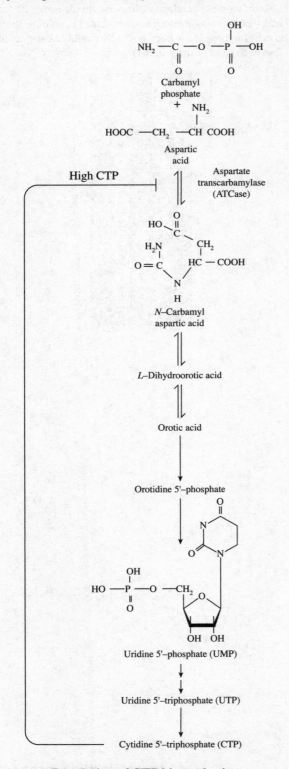

Regulation of CTP biosynthesis

61. Which of the following is true when the level of CTP is low in a cell?

   (A) CTP is converted to ATCase.
   (B) The metabolic traffic down the pathway increases.
   (C) ATCase is inhibited, which slows down CTP synthesis.
   (D) The final product of the pathway is reduced.

62. This enzymatic phenomenon is an example of

   (A) transcription
   (B) feedback inhibition
   (C) dehydration synthesis
   (D) photosynthesis

63. The biosynthesis of cytidine 5′-triphosphate requires

   (A) a ribose sugar, a phosphate group, and a nitrogen base
   (B) a deoxyribose sugar, a phosphate group, and a nitrogen base
   (C) a ribose sugar, phosphate groups, and a nitrogen base
   (D) a deoxyribose sugar, phosphate groups, and a nitrogen base

GO ON TO THE NEXT PAGE.

Directions: Part B consists of questions requiring numeric answers. Calculate the correct answer for each question.

64. In a diploid organism with the genotype AaBbCCDDEE, how many genetically distinct kinds of gametes would be produced?

65. Under favorable conditions, bacteria divide every 20 minutes. If a single bacterium replicated according to this condition, how many bacterial cells would one expect to find at the end of three hours?

GO ON TO THE NEXT PAGE.

66. In snapdragon plants that display intermediate dominance, the allele CR produces red flowers and CW produces white flowers. If a homozygous, red-flowered snapdragon is crossed with a homozygous, white-flowered snapdragon, what will the percentage of pink offspring be?

67. Translation is an energy-intensive process. Each tRNA that brings an amino acid costs 2 ATP to make. Every codon that binds in the ribosome costs 1 ATP. There is also a cost of 1 ATP every time the mRNA must shift to the next codon. Approximately how many ATPs are required to synthesize a protein containing 115 amino acids?

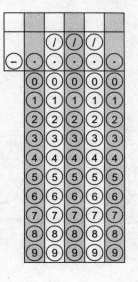

GO ON TO THE NEXT PAGE.

Question 68 refers to the following experiment.

A group of *Daphnia,* small crustaceans known as water fleas, was placed in one of three culture jars of different sizes to determine their reproductive rate. There were 100 females in the jar. The graph below shows the average number of offspring produced per female each day in each jar of pond water.

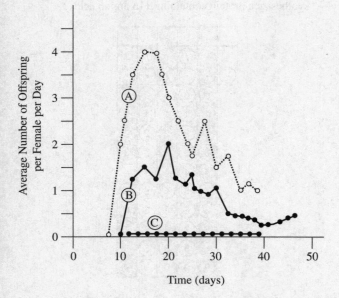

Time (days)

Key: (A) Water fleas in a 1-liter jar of pond water
(B) Water fleas in a 0.5-liter jar of pond water
(C) Water fleas in a 0.25-liter jar of pond water

68. What is the total number of offspring born in the 0.5-liter jar on the twentieth day?

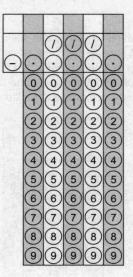

GO ON TO THE NEXT PAGE.

69. On average, there is a 90 percent reduction of energy available for each trophic level. Based on this information, 10,000 pounds of grass should be able to support how many pounds of crickets?

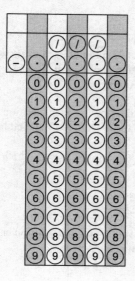

STOP

END OF SECTION I

IF YOU FINISH BEFORE TIME IS CALLED, YOU MAY CHECK YOUR WORK ON THIS SECTION. DO NOT GO ON TO SECTION II UNTIL YOU ARE TOLD TO DO SO.

BIOLOGY

SECTION II

8 Questions
Planning Time—10 minutes
Writing Time—80 minutes

Directions: Questions 1 and 2 are long free-response questions that should require about 22 minutes each to answer and are worth 10 points each. Questions 3 through 8 are short free-response questions that should require about 6 minutes each to answer. Questions 3 through 5 are worth 4 points each, and questions 6 through 8 are worth 3 points each.

Read each question carefully and completely. Write your response in the space provided following each question. Only material written in the space provided will be scored. Answers must be written out in paragraph form. Outlines, bulleted lists, or diagrams alone are not acceptable.

1. Chlorophyll is one of a class of pigments that absorb light energy in photosynthesis.
   (a) **Describe** the function of chlorophyll and what is meant by its absorption spectrum.
   (b) **Design** an experiment to investigate the absorption spectrum of a pigment given to you in a solution and make a mock graph. Using the mock results of your experiment, what color should the pigment appear as?

2. Natural selection is said to act on individuals, but evolution occurs in populations. **Explain** each process and **show** that the statement is true using an example with individuals in a population. **Identify** variation within the population. **Identify** a selective pressure. **Create** a graph illustrating the changes to the population makeup over time.

3. **Describe** the structural nature of genes. Name two types of gene mutations that could occur during replication.

4. In large multicellular organisms, signaling is necessary to maintain homeostasis. **Compare** and **contrast** the endocrine system and the nervous system with a focus on how they send signals.

5. **Describe** why fermentation is a less efficient way to produce energy than aerobic respiration.

6. **Define** analogous structures and **describe** how similar selective pressures can occur in two unrelated species.

7. **Describe** symbiosis and give an example involving humans.

8. Genes can be transferred from parent to offspring, but there are many other ways that they can be transferred, either naturally or using biotechnology. **Describe** two ways that genes can be transferred and **explain** why these types of transfers are helpful.

STOP

END OF EXAM

Practice Test 2:
Answers and
Explanations

PRACTICE TEST 2 ANSWER KEY

1.	D	24.	C	47.	B
2.	C	25.	D	48.	B
3.	D	26.	C	49.	B
4.	B	27.	D	50.	A
5.	D	28.	D	51.	C
6.	D	29.	B	52.	B
7.	C	30.	C	53.	A
8.	C	31.	C	54.	D
9.	C	32.	D	55.	D
10.	A	33.	B	56.	C
11.	D	34.	B	57.	B
12.	B	35.	B	58.	D
13.	C	36.	C	59.	C
14.	B	37.	B	60.	B
15.	D	38.	B	61.	B
16.	A	39.	B	62.	B
17.	A	40.	C	63.	C
18.	B	41.	A	64.	4
19.	D	42.	C	65.	512
20.	C	43.	B	66.	100
21.	C	44.	B	67.	460
22.	D	45.	B	68.	200
23.	A	46.	A	69.	1,000

PRACTICE TEST 2 EXPLANATIONS

Section I: Multiple Choice

1. **D** Because of the cell wall, a plant cell will not lyse when placed in a hypotonic solution. Instead, it will just swell. Both plant and animal cells perform cellular respiration (A) and have transcription factors (B) and both have vacuoles (C).

2. **C** If the contents of the cell separated from the cell wall, then water was moving out of the cell. This would cause the space between the cell wall and the cell membrane to expand, so eliminate (B). This also means that the concentration of solutes in the extracellular environment would be hypertonic with respect to the cell's interior. You can also eliminate (A): because the fluid in the cell was hypotonic to the sugar solution, fluid was moving out of the vacuole, which caused it to become smaller. Choice (C) is the correct answer because if the water was rushing out of the cell, then the water potential inside the vacuole would have been higher than outside the cell. High water potential means the water wants to move. Choice (C) says the water potential would be lower.

3. **D** This question tests your ability to associate what happens when enzymes are denatured and what would happen in the synaptic cleft. Acetylcholinesterase is an enzyme that degrades acetylcholine in the synaptic cleft. If acetylcholinesterase is denatured, acetylcholine will still be released from the presynaptic membrane and continue to diffuse across the synaptic cleft. Acetylcholine will still bind to the postsynaptic membrane because acetylcholine is not degraded. Therefore, you can eliminate (A), (B), and (C).

4. **B** The ratio of purines to pyrimidines should be constant because purines always bind with pyrimidines, no matter which ones they may be.

5. **D** Feedback inhibition occurs when something created by a process then inhibits that process. Choices (A) and (B) involve stimulations, so they can be eliminated. Choice (C) doesn't make sense because it says that "C" is an enzyme and the passage says that the enzymes are the numbers shown between the steps.

6. **D** If substance F leads to the inhibition of enzyme 3, then substances D and E and enzymes 3, 4, and 5 will be affected. The activity of enzyme 5 will be decreased, not increased.

7. **C** If enzyme 1 were inhibited, then everything that comes after that in the pathway would be reduced. A would not be reduced, but both C and X would. Choice (C) is correct.

8. **C** Based on the graph, fetal hemoglobin has a higher affinity for oxygen than maternal hemoglobin. Fetal hemoglobin does not give up oxygen more readily than maternal hemoglobin, so eliminate (A). You can also get rid of (B), as the dissociation curve of fetal hemoglobin is to the left of the maternal hemoglobin. Finally, eliminate (D) because fetal hemoglobin and maternal hemoglobin are different structurally, but you can't tell this from the graph.

9. **C** The passage says that high CO_2 and low pH lead to reduced affinity. A right shift curve would represent a reduced affinity. Photosynthesis consumes CO_2, so this would not increase the CO_2. Mitosis and digestion are not related. Fermentation is done in cells that are doing anaerobic cell respiration, so CO_2 should be released, possibly along with lactic acid, which would lower pH.

10. **A** Hemoglobin's affinity for O_2 decreases as the concentration of H^+ increases (or the pH decreases) and as the concentration of OH^- increases (or the pH increases). As the pH decreases, the affinity for oxygen will decrease, and as the pH increases, the affinity for oxygen will increase.

11. **D** The CO_2-rich environment would decrease the affinity of hemoglobin for oxygen, so the curve should shift toward the right. The regular curve seems to hit 90% saturation around 45–50 pO_2, so the CO_2-rich curve would take more oxygen to get there.

12. **B** Both prokaryotes and eukaryotes possess double-stranded DNA. Choices (A), (C), and (D) are correct differences between prokaryotes and eukaryotes.

13. **C** Bb and Bb is the correct set of genotypes that represents a cross that could produce offspring with silver fur from parents that both have brown fur. Complete a Punnett square for this question. In order for the offspring to have silver fur, both parents must have the silver allele.

14. **B** Viruses are made up of nucleic acid surrounded by a protein coat called a capsid. Don't forget that some viruses are RNA viruses and don't have a DNA genome.

15. **D** All of the choices are examples of hydrolysis except the conversion of pyruvic acid to glucose. Hydrolysis is the breaking of a covalent bond by adding water. In all of the correct examples, complex compounds are broken down to simpler compounds. The conversion of pyruvic acid to acetyl-CoA is an example of decarboxylation—a carboxyl group is removed as carbon dioxide and the 2-carbon fragment are oxidized.

16. **A** Peroxisomes catalyze reactions that produce hydrogen peroxide, ribosomes are involved in protein synthesis, and mitochondria contain enzymes involved in cellular respiration. Eliminate (B) and (D) because lysosomes are the sites of degradation; they contain hydrolytic enzymes but do not produce hydrogen peroxide. Choice (C) is incorrect, as the Golgi apparatus sorts and packages substances that are destined to be secreted out of the cell.

17. **A** The figure shows that water leaves the descending limb and flows by osmotic pressure into the surrounding space and then into the vasa recta. This means that each of those places is more hypertonic than the previous. The most hypotonic is the descending tubule (This is why the water leaves.).

18. **B** When water flows out of the tubule, it is moving by simple osmosis. However, because it is a polar molecule, it must pass through an aquaporin channel, which is facilitated diffusion.

19. **D** ADH makes the collecting duct permeable to water. If it is inhibited, then it will be unable to do that, meaning choice (D) is correct. The question doesn't say that the inhibitor is inhibiting the aquaporins directly. Instead, the inhibitor says it is inhibiting the ADH.

20. **C** To make concentrated urine, the medulla must be very osmotic because it must reclaim a lot of water from the collecting duct. To make a very osmotic medulla, the Loop of Henle must be very long. The collecting duct should be very permeable, not impermeable, because concentrated urine means that a lot of water has been reclaimed.

21. **C** The carrying capacity is the maximum number of organisms of a given species that can be maintained in a given environment. Once a population reaches its carrying capacity, the number of organisms will fluctuate around it.

22. **D** The growth of curve A is exponential, meaning that the bacteria are replicated as fast as possible without any restrictions. It looks like the populations started at the same level, and an inhibitor would not cause high growth as shown in culture A. Not being measured as often would not cause a graph like the one shown for culture B.

23. **A** Phototropism is the ability of a plant to move with regards to the sun. A climbing vine uses pressure to sense the trellis. A tree blooming once a year is using photoperiodism.

24. **C** Females with Turner's syndrome lack an X chromosome. If females with this syndrome have a high rate of hemophilia, they must not have the second X to mask the expression of the disease.

25. **D** This type of mutation is called frameshift mutation. The insertion of DNA leads to a change in the normal reading frame by one base pair. The other answer choices refer to chromosomal aberrations. Choice (A), duplication, occurs when an extra copy of a chromosome segment is introduced. Choice (B), translocation, occurs when a segment of a chromosome moves to another chromosome. Choice (C), inversion, occurs when a segment of a chromosome is inserted in the reverse orientation.

26. **C** The principal inorganic compound found in living things is water. Water is a necessary component for life.

27. **D** An extra X-chromosome could be created by a problem separating the homologous chromosomes or sister chromatids. When they are lined up, it would not be apparent that they were not separating properly, but once they start to pull apart, one could see if a mistake had occurred.

28. **D** All of the following are modes of asexual reproduction except meiosis, which is the production of gametes for sexual reproduction. Choice (A), sporulation, is a form of asexual reproduction in which spores are produced. Choice (B), fission, is the equal division of a bacterial cell. Choice (C), budding, is a form of asexual reproduction in yeasts in which small cells grow from a parent cell.

29. **B** Invertebrates lack specific immune responses including B-cells and T-cells. Phagocytes is the only answer choice that describes non-specific immune responses.

30. **C** Helicase is the enzyme that unwinds the helix. The closed helix would exist without helicase.

31. **C** The RNA polymerase is not bound, but there is a repressor bound. The transcription must be repressed.

32. **D** These four phases must be transcription. Transcription begins with a closed helix, then an enhancer binds, and then RNA polymerase is seen making RNA. Finally, the closed helix and the completed mRNA strand exist.

33. **B** Photosynthesis requires light, carbon dioxide, and water and produces oxygen and glucose.

34. **B** Avery, MacLeod, and McCarty concluded that DNA is the hereditary material because it was able to transform non-pathogenic bacteria into pathogenic bacteria. Watson, Crick, and Franklin determined the double helix, and Meselson determined that replication is semi-conservative.

35. **B** Translation, the synthesis of proteins from mRNA, occurs in the cytoplasm. DNA replication (I) occurs in the nucleus. Transcription (II), the synthesis of RNA from DNA, occurs in the nucleus.

36. **C** Crossing-over permits scientists to determine chromosome mapping. Chromosome mapping is a detailed map of all the genes on a chromosome. The frequency of crossing-over between any two alleles is proportional to the distance between them. The farther apart the two linked alleles are on a chromosome, the more often the chromosome will break between them. Crossing-over does not tell us about the chance of variation in zygotes, the rate of mutations, or whether the traits are dominant, recessive, or masked.

37. **B** In this scenario, the inside of the cell is hypertonic to the outside environment—this will cause water to move into the cell and can cause the cell to lyse.

38. **B** The most likely explanation for this phenomenon is that these birds have different ecological niches. An ecological niche is the position or function of an organism or population in its environment. Eliminate (A) because we do not know if there is a short supply of resources. Choice (C) can also be eliminated because we do not know how long the bird species live together. The breeding patterns of the bird species do not explain the lack of competition, which eliminates (D) as well.

39. **B** This relationship is an example of parasitism. Parasitism is a form of symbiosis in which one organism benefits and the other is harmed. Choice (A), commensalism, is a form of symbiosis in which one organism benefits and the other is unaffected. Choice (C), mutualism, is a form of symbiosis in which both organisms benefit. Choice (D), gravitropism, is the growth of a plant toward or away from gravity.

40. **C** If the nucleotide sequence of a DNA molecule is 5'-C-A-T-3', then the transcribed DNA strand (mRNA) would be 3'-G-U-A-5'. The nucleotide sequence of the tRNA codon would be 5'-C-A-U-3'.

41. **A** Viruses are considered an exception to the cell theory because they are not independent organisms. They can only survive by invading a host. Choice (B) is incorrect, as all viruses have genomes and some have many genes. Choice (C) is also wrong because not all viruses have a tail. You can also eliminate (D), as viruses did not evolve from ancestral protists.

42. **C** The origin of life refers to the period when the first life began. Life is shown to be present 3.5 billion years ago when there was not atmospheric oxygen present yet, so (A) can be eliminated. Also, there was no life at 4 billion years despite having CO_2 and no O_2; therefore something else must be required, and (B) can be eliminated. There were no proteins when the first life arose so (D) is incorrect. Choice (C) states that self-replicating nucleic acids were necessary, which is why life had to occur when RNA appeared.

43. **B** Functional ribosomes would not likely be found before the proteins they make. Therefore, this would be between 3.5 and 3.25 billion years ago.

44. **B** The sign that photosynthesis began was when oxygen appeared. Oxygen was absent at 3.25 billion years and present at 3 billion years so photosynthesis must have begun between those times; eliminate (A) and (C). Autotrophs perform photosynthesis not heterotrophs (they rely on eating other things for their energy).

45. **B** The similarity suggests that humans and chimpanzees are more closely related than humans and dogs. Because these two organisms share similar amino acid sequences, humans must share more recent common ancestors with chimpanzees than with dogs.

46. **A** There is no way for these two parents to produce a type O child. The only genotype that will result in type O blood is two "O" alleles, one from each parent, since the allele for O blood is recessive to the alleles for A or B blood. The AB parent has alleles for only A and B blood, so it is impossible for these two individuals to produce a child with type O blood.

47. **B** The largest amount of energy is available to producers. Population B is most likely composed of producers because they have the largest biomass.

48. **B** An increase in the number of organisms in population C would most likely lead to a decrease in the biomass of B because population B is the food source for population C. Make a pyramid based on the biomasses given. If population C increases, population B will decrease. Eliminate (A) and (C), as we

cannot necessarily predict what will happen to the biomass of populations that are above population C. Choice (D) can also be eliminated because the food source available to population C would most likely decrease, not increase.

49. **B**   Chromosomes replicate during interphase, the S phase. Choices (A) and (C) are incorrect because during G_1 and G_2, the cell makes protein and performs other metabolic duties.

50. **A**   The first sign of prophase in mammalian cells is the appearance of chromosomes, which are usually invisible. They are condensed at the start of mitosis.

51. **C**   Mitosis occurs in all of the following type of cells except mature red blood cells. Mature red blood cells are short-lived and do not divide.

52. **B**   Because neurons are not capable of dividing, it is reasonable to conclude that these cells will not complete the G_1 phase. This is a reading comprehension question. The passage states that cells that do not divide are arrested at the G_1 phase. Choice (A) is incorrect because these cells will not be committed to go through cell division. You can also eliminate (C) and (D), as the cells will not enter the M or S phase.

53. **A**   According to the graph, the resting membrane potential of the muscle fiber is close to –90mV.

54. **D**   Refer to the second graph about the membrane permeability of ions during a muscle contraction. During depolarization, the membrane is permeable to Na^+. Choice (A) can be eliminated because the voltage-gated K^+ channels do not open until after depolarization. Eliminate (B) because the concentration of Ca^{2+} does not become more negative. Choice (C) is also incorrect, as the membrane potential changes from –90 mV to +20 mV.

55. **D**   Refer to both graphs in the passage. In cardiac fibers, the duration of an action potential—a neuronal impulse—is approximately 0.3 seconds.

56. **C**   In cardiac muscle fibers, depolarization is prolonged compared to that in skeletal muscle fibers. Eliminate (A) because the membrane is permeable to both Na^+ and K^+. Choice (B) can also be eliminated; in cardiac muscle fibers, voltage-gated K^+ channels open during repolarization. Finally, (D) is wrong because the refractory period is longer in cardiac muscle fibers compared to skeletal muscle fibers.

57. **B**   The two most closely related organisms are the two with the most shared derived characteristics.

58. **D**   Shared derived characteristics are newly evolved traits that are shared with every group on a phylogenic tree except for one. Vertebral columns are present in every group except for the sea anemone, so it must have evolved first. Walking legs are found only in the salamander, indicating that it most likely evolved most recently.

59. **C**   Pre-zygotic barriers to reproduction are those that prevent fertilization, so you can eliminate (A), (B), and (D). Choice (C) is an example of a post-zygotic barrier to reproduction.

60. **B**   Convergent evolution occurs when two organisms that are not closely related independently evolve similar traits, such as the wings of insects and birds. Divergent evolution occurs when two closely related individuals become different over time and can lead to speciation.

61. **B**   When the level of CTP is low in a cell, the metabolic traffic down the pathway increases. This pathway is controlled by feedback inhibition. The final product of the pathway inhibits the activity of the first enzyme. When the supply of CTP is low, the pathway will continue to produce CTP.

62. **B** This enzymatic phenomenon is an example of feedback inhibition. Feedback inhibition is the metabolic regulation in which high levels of an enzymatic pathway's final product inhibit the activity of its rate-limiting enzyme. Transcription, (A), is the production of RNA from DNA. Dehydration synthesis, (C), is the formation of a covalent bond by the removal of water. In photosynthesis, (D), radiant energy is converted to chemical energy.

63. **C** The biosynthesis of cytidine 5'-triphosphate requires a nitrogenous base, three phosphates, and a sugar ribose. Pyrimidines are a class of nitrogenous bases with a single ring structure. The sugar they contain is ribose, which is shown in the pathway diagram.

64. **4** There are four genetically distinct kinds of gametes that could be produced, ABCDE, AbCDE, aBCDE, and abCDE. Notice that the only alleles that vary are A and B ($2^2 = 4$).

65. **512** If bacteria divide every 20 minutes, you would produce 512 bacterial cells. One method would be to use the equation 2^x, where x equals the number of 20-minute intervals in three hours; $2^9 = 512$.

66. **100** Because snapdragon plants display intermediate dominance, the heterozygous phenotype is affected by the alleles of both homozygotes. If a homozygous red plant is crossed with a homozygous white plant, all offspring would be heterozygous and pink.

67. **460** A 115 amino acid protein requires 115 aa × 2 ATP = 230. A total of 116 codons will bind (115 amino acids and the stop codon). A total of 114 translocations will occur. (The start codon begins in the P-spot, but there will be a translocation after the remaining 114 codons.) 230 + 116 + 114 = 460.

68. **200** The graph shows the average number of offspring per female per day. Because there are 100 females in the half liter, the total number of offspring on the 20th day would be 100 females times 2 offspring per day, which equals 200 offspring.

69. **1,000** A 90 percent reduction of productivity in the grass is a reduction of 9,000 pounds. That means the grass should be able to support 1,000 pounds of crickets.

Section II: Free Response

Sample Long-Form Free Response 1

Chlorophyll is the principal light-harnessing pigment found in the thylakoid membrane of chloroplasts. Chlorophyll comes in two main types, chlorophyll *a* and chlorophyll *b*. Each type has a different wavelength absorption spectrum. When chlorophyll absorbs light, electrons within it are energized. Photosynthesis occurs when sunlight activates chlorophyll by exciting its electrons. Chlorophyll reflects green light, which is why plants appear green.

To determine the absorption spectrum of an unknown pigment, the pigment would have to be exposed to many different wavelengths of light and the amount of light absorbed at each should be measured. This can be done with a spectrophotometer. The wavelength of the entering light can be varied to see which wavelength is most absorbed by the solution. The least absorbed wavelengths are reflected back and that determines the color we see. As the graph shows, it would be green.

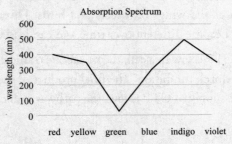

Sample Long-Form Free Response 2

Natural selection is the superior reproduction of individuals with certain inheritable traits. Those traits will contribute to the next generation, and less desirable traits will not be "selected" or passed along to the next generation. This cannot happen to a population because populations do not reproduce.

Evolution is the change of the gene pool of a population over time as certain traits disappear and others are selected. This cannot happen to an individual because an individual will have the same DNA throughout its lifetime.

As an example, let's look at a population of squirrels. Half of them are dark brown and half of them are grey. The grey squirrels are a very similar color to the pavement of the road, cause them to blend in. Motorists often hit the grey squirrels on accident because they cannot see them until it is too late. The brown squirrels are very easy to see against the grey pavement, and the motorists slow down and drive around the squirrels. Over time, the population of squirrels changes and more and more brown squirrels are seen. This is because the brown is selected because it has an advantage over the grey squirrels (less likely to be killed by a motorist). Each squirrel is still either brown or grey, but the population overall is changing. The selection is at the squirrel level, but the evolution is at the population level.

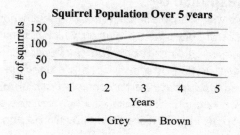

Sample Short-Form Free Response 3

A gene is a heredity unit located at a specific locus along a chromosome. Genes are made up of DNA, and DNA is made up of repeating subunits of nucleotides. A nucleotide consists of three parts, a five-carbon sugar, a phosphate group, and a nitrogenous base.

A gene mutation is a change in the sequence of base pairs in a DNA molecule. It results from defects in the sequence of bases. This can be caused by an error of the DNA polymerase or because of some other type of DNA damage. Mutations that involve a single base change in the DNA sequence are called point mutations. This type of a mutation can cause a change in the mRNA codon, which might cause a change in the sequence of the polypeptide. These are base substitutions involving a single DNA nucleotide being replaced by another nucleotide. Another type of mutation is a frameshift mutation, caused by an insertion or deletion of bases in the DNA sequence. This causes a shift in the reading frame of DNA. This often causes major changes in the sequence of the poylpeptide because many codons can be disrupted.

Sample Short-Form Free Response 4

The endocrine system works when glands release messengers called hormones into the bloodstream. The hormones can be either peptide based or steroid based. They travel to a particular area of the body and then they bind to either an extracellular receptor (peptide hormones) or an intracellular receptor (steroid hormones). This causes changes in the cell, which causes the desired effect of the hormone.

The nervous system is a system of neuron pathways that go throughout the body. Signals are sent from out in the body and into the central nervous system (brain and spinal cord). The message format of neurons is action potentials, which are waves of positive charges that travel down neuron axons. Neuron axons synapse with other neurons, and neurotransmitters are sent across the synapse to the next neuron or effector cell. Certain types of cells, like muscle cells, will contract when they are stimulated by a neuron signal.

The similarity is that each system is responding to a stimulus and passing a message to another place in the body. The nervous system uses action potentials traveling down neuron circuitry, but the endocrine system uses hormones traveling through the bloodstream. The nervous system is also much faster acting than the endocrine system.

Sample Short-Form Free Response 5

Fermentation occurs when oxygen is not available to act as the final electron acceptor. When this does not occur, then NADH and $FADH_2$ are not able to be regenerated into NAD^+ and FAD, respectively. Instead, the only step in respiration that is occurring is glycolysis, which makes 2 ATP, 2 NADH, and pyruvate. Pyruvate can act as the final electron acceptor and allow NAD^+ to be regenerated from the NADH made in glycolysis, but the only energy that has been produced is the net of 2 ATP. Since not all the energy can be mined from the carbon when processed in fermentation, the process is less efficient. Additionally, the process of reducing pyruvate makes lactic acid or ethanol, which are toxic in large amounts. Aerobic respiration is better.

Sample Short-Form Free Response 6

Analogous structures are ones which have similar functions, but whose underlying structures are very different. They are often the product of convergent evolution, which is when differing organisms end up in the same environment and thus are subjected to the same forces of natural selection. An example of analogous structures is the wing of bird and the wing of a bee. Both are used for flight, but their components are completely different. For those same adaptations to occur in both unrelated species, the same selective pressures must have occurred. For example, the ability to fly may give animals access to an unoccupied niche. It might also be a good way to avoid predators. So, in both birds and bees there was a benefit to flying.

Sample Short-Form Free Response 7

Symbiosis describes a situation in which two organisms co-exist. If the host provides a benefit, but remains unimpacted by the other organism, then the relationship is commensalistic, but if both organisms benefit, then it is mutualistic. An example involving humans is the bacterial flora of the skin or of the intestine. Many of these bacteria provide a protective or digestive benefit to humans, while some harmlessly utilize the body for food. Parasitic relationships are another type of symbiosis, but in this instance one individual is harmed in the process.

Sample Short-Form Free Response 8

In bacteria, genes are passed by conjugation between nearby bacteria. This is not a type of reproduction, but it is just gene swapping. A plasmid is passed between two bacteria, who then go their separate ways.

Viruses can also pass genetic information around, especially to bacteria. When a virus integrates inside a bacterial cell genome, it can leave a bit of genetic material and take some with it on accident. These bits join the bacterial genome. This process is called transduction.

These processes are helpful because they increase the genetic variation among the population. Bacteria replicate by fission, which is basically just making copies of themselves. Conjugation mixes up the gene pool a bit. Transduction does the same thing by picking up and dropping bits of DNA here and there. More genetic variation is better for survival of a population.